Humankind
at the Brink

By the same author

Homo Sapiens in Decline (1973)

Human Developments from an African Ancestry
(Karnak House, London, 1990)

Has Hawking Erred (Janus, London, 1993)

Physics or Metaphysics (Janus, London, 1998)

Humankind at the Brink

*The need to re-interpret
human history –
shedding new light on
the diffusion controversy*

Gerhard Kraus

JANUS PUBLISHING COMPANY
London, England

First published in Great Britain 1998
by Janus Publishing Company Limited,
Edinburgh House, 19 Nassau Street,
London W1N 7RE

www.januspublishing.co.uk

ISBN 1 85756 304 2

Phototypeset in 10.5 on 12 Baskerville
by Keyboard Services, Luton, Beds

Cover design Nick Eagleton

Printed and bound in Great Britain by
Antony Rowe Ltd, Chippenham, Wiltshire

Dedication

This work is dedicated to the late Professor Raymond A. Dart (1893–1988), famous for the discovery of *Australopithecus*, an event basic in the history of Human Evolutionary Studies. Dart's view that Africa was the cradle of humankind, now gradually accepted, took a long time to make headway against the Asia school of thought.

Portrait of Raymond A. Dart and the Taung child, painted by Alma Flynn of
Johannesburg in 1983, on the occasion of Professor Dart's 90th birthday

Commenting on a previous draft of the first part of the present work, Professor Dart wrote: 'I have gone through your manuscript, which I found deeply interesting and even enthralling. I had no idea that anyone had the breadth and vision you display. I trust that there are a sufficient number of people in both Europe and North America who are widely educated and sympathetic to your fundamental theses. What a magnificent understanding you have for humanity's past.'

Contents

HUMANKIND AT THE BRINK

Illustrations

Acknowledgments

Thanks are due to the following personalities for their encouraging comments and constructive criticisms during the several years of preparation preceding the publication of this volume:

Dr Jonathan Benthall, Director, Royal Anthropological Institute, London

Prof. Allan Bilsborough, Dept of Anatomy, University of Durham

Prof. Emeritus, George F. Carter, Dept of Geosciences, Texas A&M University, USA

The late Prof. Raymond Dart, Dept of Anatomy, University of the Witwatersrand, Johannesburg, SA

Prof. Michael Day, prominent British anatomist, St Thomas Hospital, London

Prof. Ainsworth Harrison, Dept Biological-Anthropology, University of Oxford

Dr Jacquetta Hawkes, archaeologist

The late Prof. Edmund Leach, Dept of Social Anthropology, University of Cambridge

James Mellaart, Dept of West Asian Studies, Institute of Archaeology, London

Dr P. R. S. Moorey, archaeologist, Ashmolean Museum, Oxford

Prof. Emeritus, John L. Sorenson, Dept of Anthropology, Brigham Young University, Utah, USA

Dr Helen Whitehouse, archaeologist, Ashmolean Museum, Oxford

Prof. Bernard Wood, Dept of Anthropology, University of Liverpool

Acknowledgments

Prof. Colin Turnbull, Dept of Anthropology, Vassar College,
USA

Prof. Fred Wendorf, Dept of Anthropology, South Methodist
University, Texas, USA

Dr Werner Kaiser, Egyptologist, German Archaeology Institute,
Cairo, Egypt

Dr G. Dreyer, Egyptologist, German Archaeology Institute,
Cairo, Egypt

Dr Stephen Quirke, Egyptologist, Egypt Dept, British Museum,
London

Special thanks are due to Professor Phillip V. Tobias, of
Witwatersrand University, South Africa, (one of the world's
leading authorities on brain evolution and paleo-anthropology)
for his continuous support and encouragement during the preparation of this book. Other close supporters include Joan Covey,
author on Anthropology, California, USA; Tertius Chandler,
anthropologist, Berkeley, California; Rafique A. Jairazbhoy,
historian, discoverer of extensive Egyptian links between Egypt
and pre-Columbian America presently, Karachi, Pakistan; and
the late Clayton E. Joel, anthropologist and twelve years editor
for Historical Diffusionism (formerly the New Diffusionist),
Potton.

Introduction

Establishing the true facts to human history is of decisive importance for contemporary society. On it depends not only the survival of modern civilisation but the survival of humanity as a whole.

One of the primary goals elucidated in this book is to expose the many misconceptions which plague most of the teachings of contemporary anthropology and which more or less amount to a misinterpretation of the historical process. One of its most blatant examples is the 'fallacy of cultural evolutionism'. It is to its discussion and exposure that the main body of this book is devoted; leading directly to the alternative, a 'diffusionist interpretation' of human cultural history.

Why then the title 'Humankind at the Brink? The reason is that now after four million years of hominid evolutionary development, with *Homo sapiens* emerging as the sole hominid survivor, humankind's cultural development (including science and technology) has reached a point where the creation of nuclear explosives (so far remaining globally uncontrolled) has put us on the brink of total annihilation.

Part One

Hominid Evolution and Cultural History

1

Questioning the Concept of Cultural Evolution

For many decades cultural and social anthropology has been dominated by the doctrine of 'cultural evolutionism'. This book reveals that this concept can no longer be sustained.

The idea of cultural evolutionism dates back to German theorist Adolf Bastian, becoming further developed by such luminaries of the genre as Herbert Spencer, James Frazer and Levi-Strauss. Finally, cultural evolutionism finds its rationale in the psychology of Sigmund Freud, who rooted it firmly in Lamarckian inheritance.

Many knowledgeable readers will express surprise when being made aware of this Lamarckian linkage. Yet there is proof that Freud was a convinced Lamarckist. In a letter to one of his associates, Sander Ferenzi, Freud wrote: 'Our intention is to place Lamarckism entirely on our basis and show that need which creates and transforms organs is nothing other than the power of unconscious ideas over the body.'

The fact that Lamarckism, due to its incompatibility with modern genetical science, has become outdated, removes the foundation on which the edifice of cultural evolutionism is ultimately built. The consequent re-assessment of hominid biological evolution plus a re-evaluation of hunter-gatherer studies reveals that cultural evolutionary doctrine can no longer be accepted.

Thus, instead of a regular evolutionary linked culture process pervading human history, we find a medley of irregularity occurring cultural events spanning the ages. And within this revised scenario, cultural parallels, even if separated by wide geographical and time gaps, are found to be inter-related and, as a rule, can be traced to common origins.

Based on these considerations, the present treatise offers a long overdue critique of strongly entrenched (though now obsolete) ideas held by many contemporary cultural theorists, who unaware of the Lamarckian linkage of these ideas, still persist in embracing the cultural evolutionary theories linked to Freudian psychology.

The origins of the belief in cultural evolutionism can be traced back to events which followed Columbus's discovery of America. Returning explorers, missionaries and adventurers who swarmed the world brought back a flood of accounts relating to newly discovered lands and their populations. There followed a more systematic exploration when governments and learned societies sent out geographers, naturalists and ethnologists. The latter included Alexander von Humboldt and Charles Darwin, and in more recent times the Seligmans, Radcliffe-Brown and Malinowski, also W. H. R. Rivers, A. C. Haddon, and A. M. Hocart, to mention just a few; while armchair ethnologists, such as James Frazer, catalogued and analysed the collected material. One astonishing discovery thereby noted was that many of the newly observed strange customs, institutions, and beliefs had identical counterparts in widely dispersed areas of the world, a fact amply documented among others by Frazer in *The Golden Bough*.

How did these ethnological discoveries become linked with concepts of evolution? It is suggested that the original impetus emanated from Darwin's work *The Origin of Species*, which was the result of the observations he made during his voyage aboard the *Beagle* in his capacity as naturalist. Darwin's *The Origin of Species* became subsequently the catalyst for biological evolutionary ideas during the nineteenth century. Darwin supplied the framework wherein all species of organic life could be related to a common ancestry. Within this biological universe, the transmutation of living forms was perpetrated by natural selection. Darwin had a predecessor in Lamarck, who attributed such changes (or transmutations) to a mode of inheritance based on the transmission of physiological and psychological characteristics acquired during lifetime and affected by the greater or lesser use of organs and the direct influence of external conditions on the organism.

In fairness to Lamarck, whose theory of heredity has incidentally largely been discarded, Darwin himself adopted part of Lamarck's inheritance theory, and even attempted to rationalise it in his theory of 'pangenesis'. Despite such diversions Darwin's biological evolutionary views remained principally based on natural selection with minor concessions to Lamarckian heredity. In contrast, Lamarck's theory omitted all notions of natural selection, since at his time they were scarcely known. Lamarck published his theory in 1809, the year Darwin was born.

Also Darwin showed only passing interest in ethnological science, his main efforts being devoted to the universal process of organic life in which *Homo sapiens* played only a marginal role. When Alfred Russel Wallace questioned Darwin as to whether he would discuss 'man' in the forthcoming *Origin*, Darwin replied: 'I think I shall avoid the subject as ... surrounded by prejudice. Though I fully admit ... it is the highest and most interesting problem for the naturalist.* Lamarck and his followers on the other hand devoted themselves more to *Homo sapiens* and his cultural life. Thus Adolf Bastian, an ardent follower of Lamarck (a German scholar born in 1826 when Darwin was 17 years old), argued that by a general law, the psychic unity of man everywhere produced similar ideas ... This led in turn to the belief in the

---

* From *The Life and Letters of Charles Darwin,* Francis Darwin Editions, John Murray, 1888, vol. 2, p. 6.

independent evolution of culture. Glyn Daniel (1994:91) described Bastian's theory as a form of 'super-organic, or cultural, or social evolution'.

Herbert Spencer's (a contemporary of Darwin) ideas on the subject were similar to Bastian's though more specific. Freeman (1974:216), has pointed out that 'by 1873 Spencer had systematically extended his fervently held Lamarckian beliefs to human social evolution. Spencer had advanced the theory that the mental and social evolution of the species *Homo sapiens* was primarily caused by the inheritance of acquired characteristics producing gradual and inevitable modifications of human nature and human institutions.'

Spencer was not alone in holding Lamarckian views; even Darwin subscribed to them to a degree, and both were later joined by such authorities as E. B. Tylor, Lewis Morgan, James Frazer, and others. They developed the theory of 'unilinear cultural evolution' as the best possible explanation available at the time for the presence of identical cultural traits paralleled in different parts of the world.

T. I. Dyer (1890:247), wrote in *Nature*: 'Darwin's difficulty was exactly that of everyone else's, for, in the absence of any knowledge of the mechanism of genetics, the inheritance in certain instances, of the effects of the "use and disuse of parts" and of the "direct action of external conditions" seemed plausible enough and was accepted by virtually all the leading biologists of the 1870s.' Even Weismann (1882) still gave some credence to the 'transforming influence' of 'direct action'.

Unilinealism in culture arose from the idea that the cultural evolution of *Homo sapiens* was a continuation of his biological evolution. In unilinear cultural evolution it is assumed that cultural patterns in different parts of the world are genetically unrelated and yet pass through parallel sequences. E. B. Tylor, for example, proposed an evolutionary cultural line of human progress uniformly applicable throughout the world passing through the stages of savagery and barbarism to civilisation (*Anthropology*, 1881); while Lewis Morgan spoke of an inevitable cultural evolution passing through seven different stages –

leading separately in Egypt, China, and Middle America, to the civilisations of literate cities (Daniel, 1964:68).

At a later stage Freud, Jung, and other psychologists thought in similar cultural patterns. According to Freud (following Frazer): 'In an evolutionary sequence magic is succeeded by religion in which man surrenders part of his powers to super-natural beings, and this in turn is succeeded by science.' Freud's biographer and pupil Ernst Jones relates (1991:296) that – like Spencer – Freud was an ardent Lamarckist: 'Freud early cherished the Lamarckian belief to which he had adhered throughout his life.' And in a letter to Ferenczi (1961:442), Freud remarked: 'Our intention is to place Lamarckism entirely on our basis and show that "need" which created and trans-formed organs is nothing other than the power of unconscious ideas over the body . . . in short, the "omnipotence of thought.' Another Freud biographer, Wolheim (1971:219) remarked that Freud himself did not articulate any coherent social theory. Yet contemporary ethnologists (among them Levi-Strauss) have eagerly formulated ethnological theories related to Freudian and Jungian tenets, despite the latter's Lamarckian tendencies which have long since been discarded in the light of genetic science. Thus it can no longer be maintained that there is a psychological mechanism in the human makeup based on the genes which explains how Lamarckian inheritance can be produced.

Jung's *Analytical Psychology* (Macmillan: 983), has been criticised as metaphysical and unscientific. According to Jung, 'there is a deeper level (of the personality), namely the "collective unconscious", this being the collective beliefs and myths of the race to which individuals belong, termed "archetypes". On the deepest level some of these will be "universal archetypes" common to all humans.'

The prevailing view held by most modern biologists about Lamarckian inheritance has been competently summarised by Julian Huxley (1957:35):

With the knowledge which has been amassed since Darwin's time it is no longer possible to believe that evolution is brought about through so called inheritance of acquired characters-the direct effect of use or disuse of organs, or of changes in the environment, or by the conscious will of organisms, or through the operation of some mysterious vital force, or by any other inherent tendency. What this means in terms of biology is that all the theories lumped together under the heads of orthogenesis and Lamarckism are invalidated. They are ruled out: they are no longer consistent with the facts. Indeed, in the light of modern discoveries they no longer deserve to be called a scientific theory but more as speculation or even superstition disguised in modern dress.

Even the late Arthur Koestler, an ardent Lamarckist, admitted that Lamarckism has never been able to provide a physiological explanation for the inheritance of acquired characteristics (Janus, 1978:273).

Yet in spite of the eclipse of Lamarckism, and later the rejection of unilinear cultural evolution (now dismissed as obsolete by a majority of contemporary anthropologists), it can be shown that the tenets of both Lamarckism and cultural unilinealism continue to hound contemporary ethnological and social anthropological thought under such labels as 'multilinear cultural evolutionism', 'functionalism' and 'structuralism'. If closely analysed, almost all these theories have basically remained either directly or indirectly Lamarck orientated, as are most other theories linked to the concept of cultural evolution. Why? Since they cannot be explained in terms of natural selection and genetic inheritance, they must in the last resort depend on Lamarckian factors to justify their existence.

On the other hand, the dependency of biological evolutionary concepts on natural selection as the final arbiter is now firmly established in modern biological evolutionary theory and finds expression in Julian Huxley's dictum (1957:75) that: 'No evolutionary trend can be maintained except by natural selection, and natural selection can only work on what is biologically useful to its possessors.'

Darwin's term 'natural selection' (although maintained for

convenience sake) as he admitted himself (1958:82), 'can be seen as a misnomer', because it actually describes an act of 'natural elimination'. Darwin came upon his 'theory of natural selection' by first observing how animal and plant breeders selected chosen specimen for further breeding, thereby improving the race. The idea followed that a similar process might occur in nature, and so in order to distinguish the natural process from deliberate selection he called it 'natural selection'. Darwin reasoned that while human manipulation was a deliberate act, nature's process, was not selective but eliminative. Under the basic conditions of nature – including the Malthusian principle of proliferation of progeny and the subsequent struggle for existence – the lesser adapted specimen will be eliminated while the better adapted, or fitter ones will survive. This eliminative aspect must be kept in mind whenever the term 'natural selection is used. Sexual selection in animals as elucidated by Darwin is perhaps a borderline case.

The process of biological evolution has been described by Darwin as 'descent by modification'. Darwin himself used the word 'evolution', a term actually originated by Spencer, only in his later years. It was Spencer and not Darwin who popularised the term 'evolution', using it for the first time in an article entitled 'The Ultimate Laws of Physiology' in 1857. It was also Spencer and not Darwin who coined the phrase 'survival of the fittest in his *Principles of Biology* (1866:444, originally published 1864). One can now expand on this in the light of post-Darwinian genetic science (i.e. Neo-Darwinism), and propose that biological evolution is the continuous diversification of living forms perpetrated by genetic inheritance under the influence of natural selection. It involves no directional or necessarily progressive principle.

Consequently, the term 'evolution' when applied to cultural or social processes is misplaced, and can only be rationally used in a biological sequence, based on genetics and natural selection. It is significant that one of the leading anthropologists of our day, Levi-Strauss (1968:3) admitted the above derivation of this idea by saying: 'The evolutionist interpretation in anthropology clearly derives from evolutionism in biology.' In the face of this significant admission most contemporary social anthropologists have remained blinkered to this truth.

What else can we learn from contemporaneous researches into history? Above all they show that most of contemporary anthropology by ignoring the historical diffusionist approach to culture, has continued to cling to the obsolete cultural evolutionary theory. This fact has been pilloried by the prominent American scholar Bartauld Laufer as long as about 80 years ago. In a review of Lowie's 'Culture and Ethnology' (*American Anthropology*, 1918, 20:87/8), Laufer wrote:

The theory of cultural evolution, to my mind, is the most inane, sterile and pernicious theory ever conceived in the history of science (a cheap toy for the amusement of the big children) ... culture cannot be forced into a straight-jacket of any theory whatever it may be, nor can it be reduced to chemical and mathematical formulae. All that the practical investigator can hope for, at least for the present is to study each cultural phenomenon as exactly as possible in its geographical distribution, its historical development and its relation or association with kindred ideas. [In other words: 'diffusion', G.K.]

2

The Fossil Record

Ever since hominid fossils were found and described (or dismissed as non-hominid), controversy has raged about the correct succession of hominid descent. It took Professor Raymond Dart's Taung Baby, *Australopithecus africanus*, several decades to be confirmed as a genuine hominid. Meanwhile disputes about the exact succession of hominid species have involved such types as *Australopithecus boisei, Australopithecus afarensis, Australopithecus robustus, Homo habilis, Homo erectus* and *Homo neanderthalensis*. Formerly it was thought that all of these may have had a common ancestor, the pre-hominid *Ramapithecus (R)*. This theory has now been abandoned with new discoveries of more complete specimens of *R*. and a re-evaluation of their morphology, indicating that *R*. may after all be in the orang-utang line, which split off from the African hominid lineage between about 16 million years and 10 million years BP (P. V. Tobias, 1983b.).

More recent research proposes drastic adjustments in the fossil record concerning both Neanderthalers and modern *Homo sapiens*. Hitherto one theory had been that *Homo sapiens* evolved from the Neanderthalers, who emerged about 75,000 years ago in the evolutionary time scale. At that stage Cro-Magnon, the modern type of *Homo sapiens*, were presumed to have appeared about 40,000 years ago. Now the Neanderthalers have been reassigned a time sequence of between 150,000 and

30,000 years BP (Gowlett, 1986), being classified as *Homo sapiens Neanderthalensis*.

Doubts about the Neanders being a subspecies of *Homo Sapiens* were published in the 4 July 1997 issue of the US journal *Cell*. A research team under Savante Paabo at Munich University carrying out tests on the famous Neanderthal skeleton, found that its DNA differed in many respects from the DNA of modern humans.

The experiment involved DNA samples from 2,000 people in Africa, Asia, Europe, Australia, Oceania and North America. It was established that *Homo sapiens* specimens differ from each other by an average of eight variations while the Neander specimen differed in 27 respects from modern *Homo sapiens*. In comparing contemporary people with chimpanzees the difference is 55. The overall results suggest that the Neanders diverged about 600,000 years ago from the line that would eventually become today's *Homo sapiens*

The emergence of the modern *Homo sapiens* is now put at between 120,000 and 100,000 years ago, while what is called the more archaic type of *Homo sapiens* reaches back to between 500,000 and 300,000 years ago (having arisen from *Homo erectus* stock). Referring to the evolutionary transmission from *H. erectus* to *H. sapiens* for which he allocates a space of 500,000 years, P. V. Tobias remarks (1985b), 'This phase of hominidisation is bedevilled with so many problems of definition and of morphological appraisal that estimates of dating vary from 0.75 to 0.25 million years BP.'

The two rows of diagrams that follow illustrate differences of opinion on the succession of hominid descent, while the distances plotted bear no relation to chronology.

By analysing a DNA segment (a form called Mitrochondial DNA, which is inherited only through the female) scientists in 1987 established the 'Eve' hypothesis, maintaining that all present humans (*Homo sapiens sapiens*) were descended from a single female who lived in Africa about 200,000 years ago. Latest researches By R. L. Dorit, Yale, and Walter Gilbert, Harvard (published in their journal *Science*, May, 1995), have now rejected the controversial 'Eve' hypothesis. They found

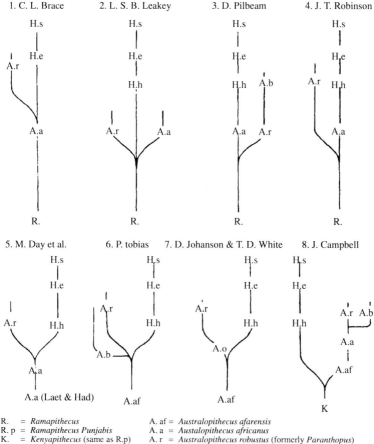

R.   = *Ramapithecus*
R. p = *Ramapithecus Punjabis*
K.   = *Kenyapithecus* (same as R.p)
H. h = *Homo habilis*
H. e = *Homo erectus*
H. s = *Homo sapiens*

A. af = *Australopithecus afarensis*
A. a = *Austalopithecus africanus*
A. r = *Australopithecus robustus* (formerly *Paranthopus*)
A. b = *Australopithecus boisei* (formerly *Zinjanthropus boisei*)

1 C. L. Brace presents one continuous line of descent with A.af being included in A.a (from 'Lucy' 1981).
2 L. S. B. Leakey assumes a common ancestor R., leading straight on to H.h. with an early separation of both, A.r and A.a.
3 D. Pilbeam projects a direct line from R. to A.a etc., with a side branch incorporating both, A.r and A.b.
4 J. T. Robinson offers the same as 3 except that the branch-off accommodates only A. r (quotes 2, 3, 4) from *Encyclopedia of Anthropology*, 1976.
5 M. Day *et al.*, does not recognise A.af, which he considers part to the single species A.a; with A.r branching-off separately.
6 P. V. Tobias treats A.a as the rootstock of all other subsequent hominid; allocates A.r and A.b to separate branches (1983a).
7 D. Johanson and T. D. White, treat A.af as the rootstock of all other hominids, allocates A.a and A.r to a separate branch.
8 J. Campbell (1983), projects K. as a common ancestor, leading directly to H.h, etc.; with a separated branch accommodating A.af, A.a, A.r, and A.b.

that there is absolutely no indication that one male or female, or even a single couple, were the sole ancestors of modern *Homo sapiens sapiens*. They tested the Y chromosomes of males from every major racial and geographical region of our globe. Their conclusion: Everyone living today is descended from a small population of a few thousand people who may have lived 270,000 years ago.

3

Brain Evolution in General

Present assessments of hominid brain evolution must necessarily be based on brain size only, since no other valid denominator exists to measure the functional value of the brain. Neither is there any discoverable link between brain size and intellectual potential, though, as will be hinted later, a certain marginal relationship seems to exist.

Dean Falk (1980), has pointed out that it is largely through R. L. Holloway's efforts that we know that brain size increased from *c*. 450 cc. in the Australopithecines to *c*. 1,400 cc. in modern humans, and that hominid brain size has more than tripled since the time of *Australopithecus* (Holloway, 1975). As to the relationship between brain size and body size, Holloway and Post (1980) correctly point out that relative brain size has increased steadily from *Australopithecus* to *H. sapiens*. Passingham (1975) points out: 'Brain size in Homo sapiens is *c*. 3.1 times that predicted for non-human primates of equivalent body size. According to Falk:

> Further, indices show that man's neocortex is *c*. 3 times as large as that expected for a non-human primate of the same body size. It appears, that encephalization at birth is not higher in humans than in non-human primates ... the postponement of three-quarters of brain-growth in

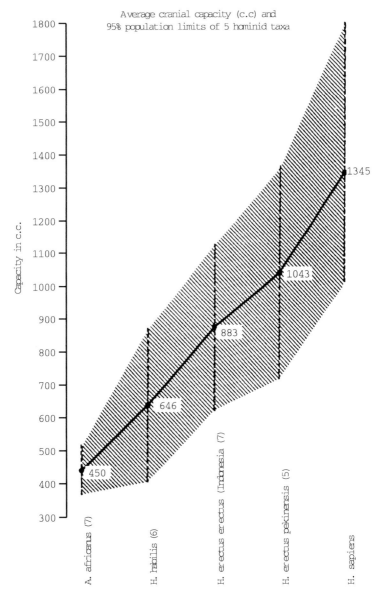

Average cranial capacity (c.c) and
95% population limits of 5 hominid taxa

The above table prepared by P. V. Tobias (who is a close collaborator of Professor Holloway), gives an up to date picture of average cranial capacities for the five taxa of hominids (1982).

humans to postnatal periods should be viewed in terms of obstetrical limitations. Hence, to pass through the pelvic canal, an human neonate must be born at a more imma-ture (more altricial) stage. Accordingly, newborn/adult brain ratios have decreased from *Australopithecus* to *Homo sapiens*.

The late Professor Edmund Leach commenting on a previous draft of mine, relating to the figure on page 16, by P. V. Tobias, had this to say: 'I appreciate that the figures are based on "average" cranial capacities but the samples are so small as to be meaningless.' This point has been duly taken into account in a separate treatise by P. V. Tobias (1983a; pp. 101–103). Remarking on the number of Early hominid individuals re-covered from African sites over three million years, P. V. Tobias admits that when the data so far collected are broken down the number of fossil individuals recovered is woefully small. 'To a human-population biologist, for whom the sample size is often a critical consideration, 485 individuals is not a par-ticularly large example. This is especially true when one con-siders that these 485 individuals spanned not less than three million years, that is, from about four to about one million years before the present. If our fossils were evenly spaced in time – which of course they are not – we should have one individual for every 6,185 – years.' P. V. Tobias concludes: 'Despite the shortcomings of the fossil hominid data, the almost explosive increase in the number of specimen in the past quarter of a century has given us a much clearer picture of the nature of the morphological hominids and of the pattern of hominid evolution.'

Past figures compiled by Professor Michael Day given on page 18 (1973) show in comparison to P. V. Tobias's table that cranial assessments of hominids have changed little during the past decades.

Average Brain Sizes

Hominids		Apes	
Homo habilis	639 cc.	Chimpanzee	395 cc.
Pithecantropus (H. erectus)	880 cc.	Orang-utang	410 cc.
Sinanthropus (H. erectus)	1,057 cc.	Gorilla	505 cc.
Modern man	1,370 cc.		
Australopithecus	500 cc.		

Other past figures compiled by Professor Bernhard Campbell (1974) are: *A Africanus* 588 cc.; *H. erectus* 950 cc, and *H. Sapiens* 1,330 cc.

The following table prepared by P. V. Tobias (1986) gives a useful summary of evolutionary trends from the ape-line to *H. sapiens*, under the heading:

Some suggested climacteric events in hominid evolution

Event	Possible/approx.dating (million years BP)
1 Orang-utang divergence	16–10
2 Gorilla divergence	10–7?
3 Hominid-chimpanzee divergence	9–5?
4 *Homo*–A. divergence	2.3 (*Homo habilis* GK)
5 Earliest stone cultural remains	2.5–2.0
6 Acquisition of speech	2.0–1.5
7 Movements of hominids from Africa to Asia	1.8–1.5
8 *Habilis–Homoerectus* transition	1.6–1.5
9 Extinction of *Australopithecus robustus*	c. 1.3
10 Acquisition of control of fire?	1.3–0.5
11 Emergence of modern human culture	0.1–0.025

P. V. Tobias (1982) suggests a span of 5 million years in which an ape-like ancestor to *Australopithecus* became converted to modern man. At an average generation length of *c*. 15 years. The following hominising trends would have become established.

Uprightness in *c*. 100,000 generations (1.5 million years), with further perfecting requiring another more than *c*.100,000 generations. Change in teeth 100,000 generations before reaching a hominid pattern and altogether 300,000 generations (4.5 million years) to reach the modern human pattern. Brain enlargement was not strikingly manifest for the first 100,000 generations and marked brain enlargement started late, at the stage of *H. habilis*, from *c*. 2.3 million years to 0.1 million years BP, i.e. in just under 150,000 generations.

Evidence further suggests that brain enlargement was one of the last hominising trends, but modern human proportions were then reached in a relatively small number of generations.

Some Basic Brain Functions

R. L. Holloway (1978, 1979), quoted by R. Leakey (1981), gives the following simplified summary:

The brain is divided into four parts, or lobes. The frontal lobe controls movement, the back or occipital lobe vision, and emotions, the side or temporal lobe, memory; above the temporal is the parietal lobe having the crucial role of comparing and integrating information that flows in through the sensory channels of vision, hearing, smell and touch. Roughly speaking, in the human brain the parietal and temporal lobes predominate, whereas in ape brains these areas are smaller.

Speech abilities are located in the Wernicke area, but the actual muscular movements for sound are located in the Broca area. Thus anatomically, the left hemisphere is rather larger than the right, and there is a detectable lump over the region that houses the Broca area. In apes the swelling is less pronounced.

R. Holloway further reminds us (quoted by R. Leakey, 1981) that the area which controls the fine action of the hands and the area governing muscular movements required for speech delivery lie very close – this may reflect shared origins.

4

Brain Size in Perspective

Various authorities have commented on the relationship of brain size and intelligence. Johanson (1982) has remarked:

> ...that the differences in brain size within our species appear to have no significant correlation with the intelligence of their owners. Rather they reflect differences in body size. Big men have big brains, but they are no smarter than small men. Men are also larger than women and have consistently larger brains, but the two sexes are of equal intelligence. Since there has always been a high degree of sexual dimorphism in hominids, it must be accepted there will be size differences in fossil skulls. If a large and a small skull are alike in every respect but size, the possibility cannot be ignored that the small one may be a female and the large one a male – and that they are the same species despite size difference.

A study of brain sizes carried out by G. Von Bonin a prominent American brain anatomist, led him, in 1963, to conclude that there was no apparent correlation between the average cranial capacity and the cultural status of the various races. Generally, the weight of the brain is a poor indicator of its functional value. Moreover, G. Von Bonin pointed out, in any major series of skulls, the capacity will vary between say, 1800 and 1200 cc., without any apparent corresponding correlation with function. Still it has been said that a certain amount of brain is necessary for normal function of the individual. Keith (1948) has spoken of a cerebral Rubicon, which he put at 800 cc.: 'Only when the cranial capacity is greater than this amount are we justified in assuming a human intelligence and that ability to learn which appears necessary for human status.'

R. Dart (1956) has disputed such arbitrary limitations by citing examples of Bantus with a capacity of 511 cc., 519 cc. and 560 cc., who functioned normally as herdboys and farmhands. 'A Bantu woman with only 340 cc. was able to do some routine work and could dance when she heard music.' Other examples by G. von Bonin are of people within between 340 and 490 cc. 'Some of these are marked as "idiots". But the fact is, that most people have much greater brain weight and that the average is about 1400 cc.'

On the other end of the scale we get the recorded brain sizes of Cromwell and Lord Byron of 2,350 cc. (Dart, 1956). Pearson (1925) showed that some very gifted persons including Leon Gambetta, Anatole France and Franz Joseph Gall, had small brains of about 1,100 grams, while Dr Johnson's was over 2,000 grams. He concluded that correlation between brain size and mental capacity was not significant.

Similar ideas have lately been voiced by Johanson. He writes (1982):

Brain size alone is now recognized as a questionable index of species identification because of its variability. People today have brains that range in size between 1,000 and 1,800 cc. and in their lower range actually overlap the brains of *H. erectus*, which run from 700 to 1,250 cc. If the largest brained *H. erectus* were to be rated against the smallest brained sapiens – and all their other attributes ignored – their species names would have to be reversed. Similarly, the habilis brain overlaps that of erectus, varying between 500 and 800 cc.

All this does not alter the fact that within the four million years of hominid evolution under review (from *Australopithecus afarensis* to contemporary *Homo sapiens*) there has been an increase in average brain size of about 1,000 cc. (from *c.* 415 to *c.* 1,350 cc.). While we therefore cannot say that increase in brain size is directly responsible for cultural progress we ought to allow it at least a marginal influence.

From all the preceding details it appears that, although in recent times a reasonable consensus has emerged, both in respect of the ages of the different hominid species, their succession, and their brain size, the situation remains still too volatile to establish a reliable lineage of hominid descent.

The Antiquity And Brain Size of Homo Sapiens

Concerning the antiquity of *Homo sapiens*, some remarkable facts can be noted. They show that during the last decades his age estimate has expanded back beyond any foreseeable bounds, while in contrast, his average brain assessment has decreased.

For example, only a few decades ago, the *Homo sapiens* span was put at not more than 10,000 years, and as recently as the late 1960s (noted while researching for my book, *Homo Sapiens in Decline* 1973), *H. sapiens*'s age estimates ranged between 60,000 and 35,000 year BP Charlton Coon (1967), put *H. sapiens*'s age at 35,000 BP; and the same figure is given by Bray/ Trump (1970). Only a few years later, Bernhard Campbell (1974), advanced *H. sapiens*'s age to between 300,000 and 250,000 BP. While Johanson (1982) put *H. sapiens* beginnings at between 500,000 BP and 200,000 BP. Finally, Joseph Campbell (1983) has referred to a fossil find at Vertescollos, Hungary, which he describes as *Homo sapiens*, with an age-tag of half a million years (this has since been corrected and reduced to *c.* 200,000 years).

As to *Homo sapien*'s brain size, W. W. Howells (1967) suggested an average value of 1,450 cc., while G. V. Bonin (1963), lists 1,400 cc. In comparison, more recent figures put the mean brain size of *H. sapiens* at 1,370 cc. (M. Day, 1973:90), 1330 cc.

(B. Campbell, 1974), 1,300 cc. (*Encyclopedia of Anthropology*, 1976) and 1,345 cc. (P. V. Tobias, 1983a).

What is also remarkable about *H. sapiens*'s brain size is, that the Upper Palaeolithic *Homo sapiens* showed bigger average brain sizes than contemporary humans. For example, G. V. Bonin (1963) wrote that the skull capacity of the Upper Palaeolithics is greater than almost all modern races by 40 cc. to 50 cc. A still greater discrepancy has been observed between the mean brain size of modern man and the classical Neanderthalers by M. Day (1973), the comparative figures being 1,470 against 1,370 – a difference of 100 cc.; Howells and Trinkaus (1980) speak of a mean for Neanders of 1,600 cc.

M. Day (1973) also quotes Thomas (1969), who stresses the shrinking of cranial capacities in recent man. Thomas estimates the male brain capacity of the Upper Palaeolithics at *c.* 1,600 cc., adding that this amounts to a mean of 1,520 cc. for the two sizes male and female taken together. He remarks that 'Since present cranial capacity is about 1,350 cc. throughout the world (personal mean: 1971), there has been thus more than 10 per cent reduction, bearing on volume and not on functional value.'

5

The Brain's Intellectual Qualities

Among contemporary studies on this subject that of Dean Falk sums up the position admirably. He writes (1980):

> We pride ourselves that human intellectual achievements are so great that they must be the result of qualitative improvement in the human nervous system. According to this line of thought, it is not just that humans have more brains than did their early ancestors, but qualitatively better brains. But is this true? Various authors have racked the fossil record of hominid brains for an answer to this question.

Below is my own effort to hint at a solution.

In my book, *Homo Sapiens in Decline* (1973), when commenting on 'The Role of Instincts', I wrote: 'The thesis of this book is that the motive of self and species preservation is the essence of all animal and human activities. In this process all biological properties have been developed by natural selection based on gene mutations. The living functions basic to this process were, in the order of their emergence, those of reflex action, instinct, and reason. While reflex actions are automatic. "Instinctive urges are the primary forces that induce animals to act purposefully (i.e., keep alive and procreate). But such action requires the aid and guidance of the senses. Sensory perceptions and their interplay form the basis of the mental process, and involve, with the aid of memory, the capacity we call thinking or reasoning." In other words, our mental powers (i.e. intelligence) are the servants of our instincts.'

The importance of the sensory input in brain evolution has been emphasised by G. von Bonin. In his book *The Evolution of the Human Brain*, (1963) he writes

> One of the most important differences between higher and lower vertebrates that we are just beginning to appreciate is the much less variegated sensory input of the lower forms and the importance of this fact for the intelligence of the animal. In this connection it should be particularly pointed out that one of the most important differences between primates and the other mammals is that the former rely more on vision than olfaction and, consequently, that the optic nerve gets much bigger and contains more fibers than in other forms. To cite but a few examples: dog and cat – about 150,000 fibers; pig and sheep – about 600,000; and monkey and man – about 1,200,000. This means of course, that there is a much richer sensory input in primates, although it certainly is not the only factor.

One may add that modern studies of the neurological structure of the brain have amply confirmed such developments. This applies particularly to the study of neuro-transmitters which transfer the electrical signals of the sensory impulses into chemical receptors, producing the desired physiological effects.

This point is further elucidated by P. V. Tobias (1971). He points out:

> An advancement of brain development has been proposed by Jerrison, producing interesting early confirmation of the rise of brain development of *Homo habilis* over the *Australopithecines*. The results obtained, expressed in billions of neurons are:

African great apes	3.4–3.6	H. erectus	5.7–8.4
Australopithecines	3.9–4.5	H. sapiens (various	
H. habilis	5.2–5.4	populations)	8.4–8.9

That there is however no definite correlation between brain size and brain quality has been emphasised by G. von Bonin (1963): '...the inside of the brain is sheathed in the "dura matter" and is seen as through a thick veil, so that very little can be discerned. It has been shown that the fissures of the human brain are quite variable and these variations do not appear to throw any light on the mental characteristics of the bearer.'

What about the other features of the brain? Certain scientists have based human brain superiority on the particular development of the frontal lobes. This view is rejected by von Bonin, who believes that the importance of the frontal lobes has often been vastly exaggerated. Also, the more reliable indices show that the frontal lobes of hominoid fossil forms differ little in size from those of modern man, except that modern man's brains are generally larger. But in both types, the indices for the various parts of the brain do not differ greatly.

These views about human brain evolution, largely accepted on the authority of von Bonin, are nowadays held by most brain anatomists. This opinion is shared by P. V. Tobias. In the Introduction to his book, *The Brain in Hominid Evolution* (1971), he speaks of the comments Dr Gerhardt von Bonin who, after a lifetime of studies of brain and endocasts concludes:

It should at least be admitted that most of what has been said and written on the sulci of the brain as they have been seen on endocasts is worth very little. A view shared by not few of those mentioned above: (i.e., Le Gros Clark, Cooper and Zuckermann 1936, Edinger 1948, Conolly 1950, Simon 1965, and Banchat and Stephan 1967).

It appears therefore that specific human brain qualities can

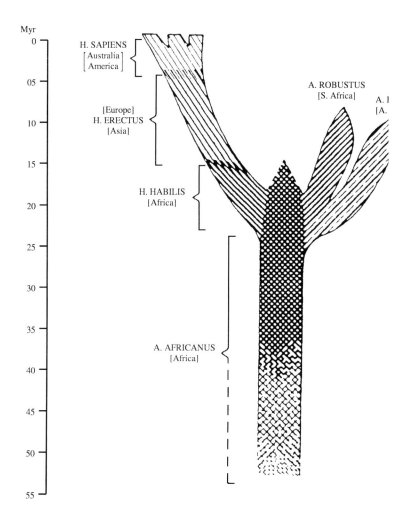

Provisional schema of hominid phylogeny showing the cladogenesis of just less than 2.5
of years ago: at that time there occurred a splitting of the hominid lineage from which flowed
a moment of explosive evolution – three different species, *H. habilis*, *A. robustus* and *A. b*
P. V. Tobias.

neither be deduced from differences in brain volume nor from the shape of the brain itself.

It is here that the factor of natural selection comes into play. Its biological evolutionary significance is expressed in Julian Huxley's dictum in 1957. 'No evolutionary trend can be maintained except by natural selection.'

P. V. Tobias, commenting on a preliminary draft of this book in a personal communication, observed that the rather condensed presentation of 'natural selection' as contained in Julian Huxley's dictum, if not expanded upon, could be misconstrued as being allied with the conventional assumption that Darwinian evolution is solely one of slow gradualism. Seen in this context it seemingly takes no account of theories of punctuated and reticulate evolution, which have lately become prominent. Reticulate (or network) evolution, as indicated by Julian Huxley (1963), specifically in humans, results from crossmating in face of differences in colour, appearance and behaviour. In other words, it resembles racial intermarriage.

In practice, however, as P. V. Tobias himself has shown, these apparent innovations are but variations of the same Darwinian principle. Thus punctuated evolution merely implies a more explosive process of evolutionary change, which could be translated as an accelerated gradualism, with the explosive event occurring at a point where variations of species branch off the main stem to form new species. According to P. V. Tobias, one such point was reached in hominid evolution *c.* 2.5 million years ago when branching (or cladogenesis) occurred from the line of *Australopithecus*, producing the separate species of *H. habilis*, *A. boisei*, and *A. robutus* (see Figure on page 27).

P. V. Tobias further elucidates this point by referring to Simpson (1953), who has suggested that at the lowest level, the process of speciation 'starts usually with minor differences between individuals, which in most local populations fluctuate from generation to generation'.

As to what may be responsible for cladogenesis, or for a moment of explosive evolution, P. V. Tobias quotes Rench (1959), who in broad terms says of this question:

The essential factor in the causation of such periods of explosive radiation is not an increase of the rate of mutation or an accumulation of macromutations, but an acceleration of differentiation, brought about by a *temporary intensification of selection* [my emphasis] due to environmental changes, e.g., by new types of vegetation or food resources, or due to colonization of new ranges with habitats unoccupied or inhabited by types inferior in competition.

Referring to the 1859 and 1866 editions of Darwin's *Origin of Species*, P. V. Tobias also points out that Darwin's concept of evolution is inherent in both punctuational and gradual evolution and what has been called recently the theory of punctuated equilibria is not in variance with Darwin's theory or the neo-Darwinian synthesis, as some have held recently, but simply represents a 'special emphasis on one part of Darwinism' (Tobias 1983a).

In the light of the above, my use of Julian Huxley's term 'natural selection' and its meaning in the context of brain evolution is justified when adding the proviso that it covers all the three types of evolutionary change as elucidated by P. V. Tobias, namely – gradualist, punctational, and reticulate evolution – with and additional emphasis of the eliminative aspect as explained below.

This arises from further comments by P. V. Tobias on my provisional draft. Referring once more to J. Huxley's categorical statement on the role of natural selection. Tobias points out that it should not be forgotten that a great deal of evolutionary change could have been non-selective in character. He also notes that Julian Huxley himself stressed this point when speaking of the large role that non-selective accidental elimination plays in evolution (my emphasis).

This argument strikes at a vital point of evolutionary theory to which I have repeatedly tried to draw attention, namely that the term 'natural selection' must be seen as a misnomer and that it describes in effect an act of 'natural elimination' (and not deliberate selection). Here I must repeat.

Darwin came upon his theory of natural selection by first observing how animal and plant breeders selected chosen specimens for further breeding and thereby improved the race. This gave him the idea that a similar process might be going on in nature. So in order to distinguish deliberate manmade selection from the random natural process, he named it 'natural selection'.

But Darwin admitted later (see 1958 reprint), that correctly applied, the term natural selection was a misnomer, because human selection is a deliberate act, while, nature's process in not selective, but eliminative. Under conditions of nature – which additionally involves the Malthusian principle of proliferation in progeny, and the subsequent struggle for existence – the lesser adapted individuals will perish, while the better adapted will survive. In other words it is 'the survival of the fitter ones'.

In view of the above, I suggest that Huxley's original dictum to become all-embracing and generally applicable to biological evolutionary processes, should be enlarged as follows: no evolutionary trend can be maintained except by natural selection – which in its true sense represents an act of natural elimination.

And, finally, Julian Huxley must have been aware of the basically eliminative aspect embodied in the term 'natural selection'. Otherwise he could not have made his original categorical and all-embracing statement.

It is this eliminative aspect we must always keep in mind when we use the term 'natural selection'.

How does this principle operate among animals in the wild, in humans, and finally in relation to human-brain quality? To take organic evolution as a whole, it can be observed (and this will later be shown in comparative death rates) that practically all animal and plant species, with the exception of modern humans and domesticated animals and plants, are subject to intensive natural selection (read natural elimination). In spite of this, however, most animal species have remained biologically constant.

Darwin (1958:66 reprint), observed that 'species will as a rule not only remain constant over long periods of time but they will also frequently remain so under changed conditions of life'. 'Certain species,' he says, 'have migrated over vast spaces and have not become greatly or at all modified.'

There are living species and even genera of animals for which evolution seems virtually to have ceased for hundreds of million of years. One existing genus of *Lingula*, according to Rhodes (1962), has a history going back 500 million years. Other groups which have persisted over long periods of time are the lungfishes; the modern crocodiles, which differ little from crocodiles of the Jurassic period; and the opossums of South America, which closely resemble the small pouched mammals that had a worldwide distribution 60 million years ago.

Soil insects such as spring-tails undergo practically no metamorphosis; they are wingless, and seem to be survivors from the epochs before insects had developed wings. An insect closely resembling them has been found in lower Devonian rocks in Scotland. It dates back some 300 million years, i.e., some 40–50 million years before the arrival of the earliest winged insects (Russel, 1966). Simson, quoted by Julian Huxley (1957), has said, 'an oyster from two hundred million years ago would look perfectly familiar to us if served in a restaurant today.'

The general principle that emerges is that normally an intensive natural selection pressure is necessary to keep animal species merely on an existing level of 'evolutionary constancy'. On the other hand, a decline in natural selection pressure is bound to lead to a reduction in biological quality.

Arriving at the hominids we can see the same principle at work in the human species. According to Gordon Childe (1966), for about 25,000 years man's bodily evolution has been virtually at a standstill. Men of the Aurignacian and Magdalenian cultures, compared with present-day men, show only negligible physiological differences, but their cultural difference is enormous. 'Progress in culture, in the human species,' he says, 'has replaced further organic evolution.'

Johanson (1982), speaking about *A. afarensis* says, 'they flourished from about four million years ago to about three million

years ago. During that time they underwent little or no evolutionary change.' David Pilbeam (1960), discussing pre-Acheulian
hominids, has pointed out that for a million years hominids
seem to have not changed much except that brain size increased
a little. W. W. Howells and E. Trinkaus (1980), refer to the
remarkably long stability in the Neanderthalers' physique.
From the time of their full establishment, c. 100,000 years to c.
40–35,000 years ago, no evolutionary change can be noticed.
Johanson (1982), describes Cro-Magnons found in southern
France c. 40,000 years old to be virtually identical with humans
of today.

While the above human examples refer mainly to periods of
a more intensive natural selection, we must ask the question
what happens to humans when natural selection pressure
becomes less intense?

A comparative study of death rates in animals and humans
supplies a clue to this question. According to Kalmus (1964),
only two out of 1,000 of some insect species will on average
reach maturity. This suffices to keep the species constant in
numbers and biologically stable. To demonstrate the intensity
of natural selection in nature, Darwin (1958), dug and cleared
a piece of ground three feet long and two feet wide where there
could be no choking influence from other plants. Then he
marked all the seedlings of common weeds as they came up,
and found that out of 375, no less than 295 were destroyed
prematurely. Darwin also observed that the winter of 1854–5
destroyed four-fifths (80 per cent) of the birds on his grounds.
He held this to be a case of tremendous destruction, considering that in human epidemics 10 per cent is an extraordinarily
severe mortality rate.

The intensity of natural selection in humans (or rather
its lack) can be deduced from modern increases in life expectancy and thus comparative death rates, as well as from
general population increases resulting from reduced death
rates.

Average life expectancies from various historical times have
been listed by A. J. Harrison (1967) as follows (see also P. V.
Tobias, 1982):

Early Bronze Age: 18 years   1789: 35$^{1}/_{2}$ years
50 BC: 22 years   1838–1854: 40.9 years
Middle Ages: 33 years   1900–1902: 49.2 years
1687–91: 33$^{1}/_{2}$ years   1946: 66.7 years

Figures showing increases in world population have been advanced by C. C. H. Cippola (1965). They read:

1650: 470 millions   1900: 1.608 millions
1750: 728 millions   1950: 2.377 millions
1800: 905 millions   1955: 2.528 millions
1850: 1.171 millions   2000: 3.727 millions
(latest estimate *c.* 6 billions)

In all societies, the largest proportion of deaths is usually due to infant mortality. In Europe, before this century, and in some less developed countries even now, of 1,000 newborn children, 200 to 500 generally die before the age of seven (Cippola, 1965). However, with preventive medicine, death rates amongst infants have been drastically reduced in most parts of the world. A. Barnett (1964), states that in England and Wales at the beginning of this century, 65 out of every 100,000 children under 15 died of diphtheria. Between the wars, with some children immunised, the figure fell to 29. In 1947, due to extensive immunisation, the death rate was down to two per 100,000.

According to the *Daily Times*, Lagos, Nigeria (14 June 1985), death rates in Nigeria dropped from 27 per thousand in the 1960s to 17 deaths per thousand in 1984. At the same time the infant mortality rate also declined from over 200 per thousand to 100 per thousand, while life expectancies increased from 35 to 55 years.

Latest available life-expectancy figures issued by the World Health Organisation are as follows: average for developing countries is now 59.7 years compared to 41 years four decades ago. During the same period life expectancy in China and East Asia rose from 42.7 to 69.9 years; and in Africa from 38 to 51.9. In South and Central America the comparable figures are 51.2 to 66 years (WHO 24 September 1989).

Historically, worldwide, there has been a continuous decrease in death rates in all age groups, running parallel with higher life expectancies and population increases, ever since *Homo sapiens* passed from the hunter and food-gathering stage of culture to one of agriculture and civilisation. This process is tantamount to a progressive reduction in the intensity of natural selection. In other words, as cultural progress intensifies and expands, natural selection de-intensifies and decreases.

Julian Huxley writes (1957):

The more elaborate social life is, the more it tends to shield individuals from the action of natural selection (i.e., the elimination of the lesser fit); and when this occurs, harmful mutations accumulate instead of being weeded out. As a result of this process, there can be no reasonable doubt that the human species today is burdened with many more deleterious mutant genes than can possibly exist in any species of wild creatures.

And I may add that such a process certainly disallows the qualitative improvement of any biological property in humans be it physical or mental.

In modern civilisation many biological deficiencies that natural selection would most likely have eliminated in former ages also survive through medical intervention. For example, as many babies are born by Caesarean section, a too narrow uterus must therefore be considered as a congenital defect. In former times, when sterile operational practice was still unknown, most pregnant women thus impeded would have died. However, today the Caesarean procedure increases the survival of many infants, who when growing up, and in turn, producing children, are likely to pass this deficiency on to future generations. Acute appendicitis, or rather the inheritable tendency to it, presents a similar case.

According to Gerald Leach, one in every 25 babies is born with physical or mental handicap and lives. Survivors may procreate in later life and pass on these defects to their offspring.

He adds: 'These figures are bound to grow as scientists uncover the mostly mysterious causes of birth defects, some of which will be genetic. In fact it is widely estimated that for every known genetic effect, there are another three or four yet to be discovered.'

The many acute genetic defects (and others with only a hereditary pre-disposition) include diabetes, various eye ailments, dental caries and malocclusin, haemophilia, mongolism, and dwarfism, gout, pre-disposition to certain cancers, pancreatic disease, gastric and duodenal ulcers, rheumatic fever, deafness, etc. Also disturbing is an alarming increase in mental illness, noticeable in recent years throughout the civilised world.

Most of the above listed defects reflects major biological disorders. However, when it comes to an examination of a possible decline in the intellectual level of an average population, we are dealing with much subtler influences, since the composite we call intelligence (or state of mind) depends on dozens of genetic determinants, each of them influencing the capacity of thinking or reasoning in a subtle and imperceptible way. A slight genetically induced imbalance will affect the brain quality of an individual; and, as such minor defects accumulate in the population, there must be a general drop in the level of intelligence.

Is there any biological mechanism, apart from natural selection, which could prevent, or even counteract such a subtle mental decline? If there is, I am not aware of its existence. Nor has any workable alternative ever been proposed, apart from failed attempts of the Galton-Eugenists of the last century – not to forget Hitlerite attempts at racial purification.

But even if there was such a possibility and natural selection could be replaced by a rigorous eugenics-orientated artificial selection or by means of genetic engineering, results would hardly be dramatic. We have shown that intensive selection in nature merely maintains the biological efficiency of a species and artificial manmade selection, if at all practicable, could not do much more.

The reader may object that the above does not present the

entire picture, and that while constancy of species is an established fact in organic life, the less frequent emergence of new variations and species is an indication that viable biological changes in organisms are possible. Indeed, we can even monitor the processes which can lead to biological change. For example, J. M. Smith (1966) has shown that the simplest genetic change in a population is the replacement at a locus of one allele by another insofar as natural selection leads to the extinction of individuals carrying the less favourable allele.

J. B. S. Haldane has calculated that a great number of selective deaths, spread over many generations, are required, before one allele has replaced another. He found that unless selection was very intense, the number of selective deaths equalled about ten to 100 times. Haldane maintained that an intensity of selection of 0.01 is a more probable figure, and if so it would take 3,000 generations of selection to replace one allele by another. Further, species probably differ by alleles at about 1,000 loci (that is, the number is probably greater than 100 and less than 10,000). If so, the evolution of a new species would take about 300,000 generations. While this figure may perhaps be exaggerated, it emphasises the great intensity of natural selection that is required to effect evolutionary changes.

We can therefore imagine the many generations needed to deliberately increase mental qualities in human beings by genetic manipulation – if such were at all possible.

What the above examples most specifically indicate is that when left to chance (i.e., in absence of any deliberate selection), as it has been left throughout the history of all culturally advancing societies, both before and since, no automatic increase in the brain's intellectual qualities can be expected. On the contrary, as we have shown, a decrease in natural selection standards, must lead to an increase of adverse mutations in a population, and thus towards a lower level of mental efficiency – or at the most optimistic assumption to a standstill in its mental level.

The causes involved become apparent when we compare Stone Age humans with their modern counterparts. Under the former conditions, all energies seem to have been concentrated on survival in an assumedly harsh environment with the aid of primitive cultural means, necessitating an alertness and intelligence certainly sharper than that required under civilisation where a sophisticated arsenal of tools, weapons, implements, and the protective embrace of modern society, with an elaborate medical science to back it, make living less severe. In addition, the store of increasing knowledge, passed on from generation to generation and incessantly added to, becomes available to most members of our civilised world, irrespective of their genetic make-up, and often replaces the need for greater intelligence. Today, the weak, the foolish, the non-alert, and the naïve, can and do survive.

In such an environment there is no selective process to prevent the mentally less endowed individuals from procreating. Consequently, the intellectual level in modern populations is much more likely to decrease than it is to increase.

One prominent critic commenting on the above has observed that he cannot see why increasing knowledge replaces the need for increased intelligence. He says it can be argued that with the great expansion in the amount and complexity of information, more, not less intelligence is needed in order to make use of it.

My response is that there are two sides of intellectual evaluation; one, the already available genetically inborn potential, and two, the possibility of making greater use of something which is possibly under-used. Hence, modern life with its greater intellectual demands in the fields of science and technology is already adequately served by making greater use of an already inborn intellectual potential. Furthermore, an increasing availability of information also eases the task. All this can be accomplished without any biologically based selective process in this direction, which, as we well know, does not operate in modern life.

To illustrate the above, I may refer to a remark, made by a South Sea Islander, to Margaret Mead: 'My father was a cannibal, but I am going to be a doctor.' Thus bridging in one generation a cultural gulf of millennia, without any genetical adjustment.

As Brace/Montague (1968) have correctly pointed out:

In literate societies, with an elaborate division of labor, the average level of intelligence in terms of pure survival value is probably lower than in cultures where the problems of survival are much more immediate. The premium placed on human intelligence in face of prolonged scarcity in the Australian desert, or at the edge of the polar ice-cap where the penalty for stupidity is death, is almost certainly greater than it is even for the most down-trodden inhabitant of Western Europe or Northern America.

The same authors point out, however, that there are mitigating factors that prevent the mental level of a population from dropping below a certain level. With the social limitations which total mental incompetence imposes on its bearers, men and women below a certain level (i.e., idiots) are unlikely to have the chance to marry and produce children.

It should now be clear to the reader that the cultural gains of early, as well as modern civilisation (including perhaps the Mezolithic and Neolithic sequences as well), have been achieved in spite of stagnating, and possibly, even of declining brain qualities.

It is significant to note that P. V. Tobias has come to similar, though not identical, conclusions. In a general assessment of human evolution and culture (*The Antiquity of Man*, 1982), he writes:

It is suggested that the main natural selective advantage flowing from brain enlargement and specially of the lower frontal, lower parietal, and upper temporal regions, was the evolution of mechanisms for the transmission of culture, and that means primarily cognitive abilities and articulate speech. By making possible a new kind of inheritance, cultural or social inheritance, articulate speech facilitated the learning of the new techniques by children of the next generation. It took the universal mammalian capacity for learned behaviour and refined it into a powerful mechanism for insuring survival. The survival of future generations is exactly what evolution is all about.

Tobias continues:

> The feedback system operated for approximately 150,000
> generations (*c.* 2.2 million years, G.K.) ... when these
> advanced stages were reached, the whole process slowed
> down. Beyond a certain point of cultural evolution [which
> should rather be read as 'cultural development', G.K.], it
> was no longer an advantage to have bigger and better
> brains. Cultural sharing and the benevolence of social life
> had taken the place of nimble wits of the individuals as an
> insurance against extinction. Encephalization (i.e. the
> enlargement of the brain) was no longer at a premium.
> One could manage and be as educable with 1,250 g of
> brain as with 2,250 g.

I conclude by citing Dean Falk (1980). Commenting on the
dramatic increase in brain size during hominid evolution, he
concludes (1980):

> Because of human technology, language and the mental
> faculties for conscious deliberate abstract thought that
> these achievements imply, *Homo sapiens* appears to be
> qualitatively more intelligent than even his 'closest' non-
> human primate relatives. Yet the search for qualitative
> brain differences that separate fossils and living hominids
> from pongids has been futile. It may be that qualitative
> differences exist at micro (e.g. neurochemical) levels and
> are not reflected in the gross paleo-neurological evidence.
> If so, the field of comparative neurology (e.g. Armstrong
> 1979, 1980) is more likely than paleo-neurology to con-
> tribute to our knowledge of qualitative human brain
> evolution.

In the meantime, Falk suggests, that at least for the hominid
brain, bigger was better.

Further, I may add, that *Homo sapiens*, the hominid with the
biggest brain, reached an average brain size of 1,350 cc. already
between 300,000 and 200,000 years ago. This is the same

average size at which modern contemporary humans (on a world scale) are assessed. Altogether this suggests a stasis in brain development which more of less continued over the entire *c.* 300,000 years of *Homo sapiens*'s existence. So far no anatomical or genetically based evidence has been produced to show that the functional, or intellectual potential of modern humans has increased during this period.

6

An Anatomy of Stone Tools

A study of the archaeological record of stone tool making reveals a striking continuity in shapes and functions extending over millions of years, beginning with the simple pebble tools of *Australopithecus afarensis* and *Homo habilis* and ending with the refined products of the Upper Palaeolithics. It further seems that successive hominid species (apart from late *Homo sapiens*) were each initially associated with a specific tool type: *A. afarensis* and *Homo habilis* with pebble-tools, *Homo erectus* with the Acheulian hand-axe complex, the Neanders with Mousterian assemblies, the Upper Palaeolithics with a blade technology, and Neolithics with polished stone tools.

The question which arises here is whether identical stages of stone tool making arose independently in different parts of the world in the form of parallel developments, or whether each technical advance had its unique place of origin, spreading from there either directly or indirectly by physical contact to other parts of the world?

As to pre-stone tool making *Australopithecines* (and some modern hunter-gatherers also), we can conjecture that they had little motivation to deliberately fashion stone tools, since there was plenty of stone debris lying about to serve as knives and scrapers without their needing retouching. In its beginnings, toolmaking may have consisted of occasionally striking some additional chips from a jagged stone to give it a more regular edge, for, as J. Gowlett observes (1984), nothing is sharper than a fresh edge.

However, such incipient tools can hardly be distinguished from unworked stone debris. Comparable is Forde's description of contemporary Semang (Malaysian hunter-gatherers). He observed (1968), that although the Semang use stone tools, these are very under-developed: 'For example, the splitting and scraping of wood is done with rough, shapeless stones, picked up at need and thrown away again.'

The oldest pebble-tools so far recorded are from Hadar Ganda, in Ethiopia, with a suggested dating of *c.* 2.6 million years. According to D. Johanson, they are comparable to tools made by *Homo habilis c.* half a million years later at Olduvai, Omo, and Lake Turkana. Pebble-tools were succeeded by hand axe-shaped implements. According to J. Gowlett (1984), 'the idea of working two opposite faces of a stone (a basic feature of the Acheulian hand-axe) can be traced right back to early Oldowan at Olduvai, where it is seen in discoids.' This development is described by Clark/Piggott (1970). The earliest assemblages comprise pebbles flaked into the form of choppers from one direction. In the second phase unidirectional flaking is still predominant, but this is supplemented by work from two directions and applied to a wider range of shapes. Phase three is marked, above all, by the predominance of bi-directional flaking. Finally, the primitive pebble forms are supplemented by others flaked on both faces, precursors of the bifacially flaked tools of the succeeding Middle Pleistocene'. From this brief account it is plain that 'evolution (read 'development') was extremely slow', since on any of the accepted chronologies it was spaced out over a period running into hundreds of thousands of years.

Nevertheless, the older form of core-tools, flakes and scrapers continued to be produced throughout the Acheulian era, and on some sites they greatly outnumbered hand-axes. The more sophisticated Acheulian type hand-axe itself, resulted from a novel technique of stone-knapping. According to R. Leakey and D. Johanson, these latter tools appeared rather suddenly *c.* $1^1/_2$

million years ago at Olduvai in association with the species *Homo erectus*. Over the following million or more years, the Acheulian hand-axe complex remained the predominant toolkit of hominids throughout the Old World quite independent from a paralleled increase in brain size.

R. Leakey has suggested that the Acheulian stone industry at Olduvai was the fountainhead from which other Acheulian sites derived their cultural capital. In R. Leakey's opinion (1981), 'The Acheulian industry continued, with steady refinements, right to the end of the sequence at Olduvai, and when some African hominids migrated north, they took their technology with them.' Acheulian hand-axe industries are known throughout the length and breadth of Africa, in the Middle East, in most of Europe and in the Indian sub-continent. Recent finds indicate an Acheulian technology in Mongolia and even in Korea (Gowlett, 1984). The latest hand-axe makers fall within the definition of *Homo sapiens*. C. Coon has written (1967):

> The most striking fact about these hand-axes is, that wherever they are found they follow the same sequence of forms. During the quarter of a million years when men made these tools, the style changed very little, but what changes were made are to be seen everywhere.' Even the most learned specialist in the archaeology of the Lower Palaeolithic, as the culture of this period is called, cannot distinguish an English hand-axe of a given age from one from Palestine or from South Africa only by variations in material and weathering.

However, as already indicated, the emergence of the hand-axe did not lead to an abandonment of pebble-tools. Almost a million years later core-tools, flakes and scrapers were still being used by *Homo erectus pekinensis*, but there were no hand-axes. In contrast, *Homo erectus javanensis*, although belonging to the same species, was found without any stone tool association whatsoever (Clark and Piggott, 1970). This latter fact supports the previous conjecture that only a limited number of hominids even within identical species made stone tools.

What has also already been noted is that, chronologically seen, the sequence of the Acheulian era attributed to *Homo erectus*, proceeded deeply into the *Homo sapiens* era. There was no profound change in the Acheulian toolkit (which commenced *c*. 1^1/$_2$ million years ago) until *c*. 150,000 BP when it was superseded by the Neanders' Levallosian-Mousterian culture complex. At that time *Homo sapiens*, had already been in existence for at least 150,000 years.

About the Mousterian (Mou.) technology, J. Gowlett cites the following details (1984):

The Mousterian toolkit is chiefly based on the use of flint-flakes about 4–7 cm long. These are shaped into specialized tools by the process of retouch. In addition to the Typical Mou., and a variant which includes hand-axes, there are the Quina and Ferrassie types which are noted for a particular sort of scrapers, and the Denticulate Mou. with its 'saw-edged' tools. In the Middle East, the Mou. emerged out of the late Acheulian sometimes by way of local variants. Also in the Middle East transitional finds between Neanders and Modern humans have been claimed for sites at Mt Carmel (Israel) and Djebel Qafzeh. Most Neander finds are associated with the Mousterian. A recent find at St Cesaire, France, a very late Neander-phase (30,000 BP) is associated with an early Upper Palaeolitic phase, the Chattelperonian. It indicates that the Upper Palaeolithic blade technology could also be mastered by the Neanders, thus indicating a transitional period. By developing this specific blade technology the Upper Palaeolithics produced an entirely new arsenal of more refined tools as well as adding implements of bones, antlers and ivory, and many other more sophisticated artefacts.

However, according to J. Gowlett (1984):

a blade technology had already appeared *c*. 100,000 years BP in tie Middle East and in Africa but never reached great favour (except in Europe, where it became dominant). In Europe the Upper Palaeolithics heralded new peaks of achievement in stone and bone working *c*. 34,000 BP, while in the Middle East a similar development is noticeable from *c*. 40,000 BP. onwards.

Of special interest among the tool-types of this period, are the Solutrean ones (*c*. 19,000 BC). They are believed to have been made over a mere 2,000 years period in France and Spain. They are manifest by the use of long narrow blades up to 30 cm long, though the usual length is 8–10 cm, (the so-called laurel-leaf points). Yet, preceding this, cruder varieties existed in the Mousterian of Central and Eastern Europe. The technique used is called pressure (or invasive flaking) and involved a very delicate craftsman-ship. What is particularly notable is that laurel-leaf points have been found during the last few thousand years in late pre-Dynastic Egypt, and more remarkably still, in pre-Columbian North and South America.

Microliths (small pointed blades), used for spear and later for arrow points, seem to have appeared as early as 90,000 years ago in South Africa. They disappeared sometime later, only to reappear again in South Africa at between 20,000 and 10,000 BP. J. Gowlett (1984), observes that it seems likely that the ideas involved in their manufacture were retained by some group, rather than reinvented.

Finally, of note are the rather crudely worked edge-ground stone axes, which appeared in an Australian-New Guinea orbit around 38,000 to *c*. 22,000 years ago. They seem to be uncon-nected with the finely ground and polished, as well as hafted, stone-axes which emerged in S. W. Asia around 7,000 to 6,000 BC. The Australian-New Guinea tool seems undoubtedly repre-sentative of one of the rare cases of independent parallel development, although resemblances are rather cursory.

What conclusion can be drawn from a review of this confus-ing welter of different stone-tool varieties, types and tech-nologies? Many tool types seem to appear quite sporadically in different parts of the world – with parallel tool forms like the

Solutrean-type laurel-leaf blades, (resulting from a most intricate technique of stone-tool knapping, known as pressure flaking), appearing in the entirely different cultural and historical environments of the Americas.

Yet out of the seeming confusion and complexity there emerges a continuity of shapes and functions stretching over millions of years. In a close study we can hardly find any complex type without some indication of it having risen from a simpler, preceding form. It also becomes obvious that truly independent parallel developments are rare, or even unproved.

Only two apparently convincing cases of the independent parallel development of stone tools come to mind, one concerns the pebble-tools of *A. afarensis*, which show no known connection with the similar Oldowan tools of *Homo habilis* dated half a million years later (*c.* 2 million years BP.). Also the former show no progression towards the Acheulian hand-axe technology, while the Oldowans do.

The other case concerns ground and polished stone axes. The first finds of fully ground and polished, hafted, axe-heads, dated between 7,000 and 6,000 BC were reported from several locations in the Fertile Crescent of South West Asia (C. Coon, 1967). Since then, examples of edge-ground (unpolished) stone axes, with some hafting evidence, have been found in an Australian-New Guinean orbit (Arnhem Land, North Australia *c.* 23,000 BC and Huon Peninsula, New Guinea, *c.* 38,000 BC). These tools resulting doubtlessly from independent developments, turned out to be roughly shaped stone axes, with only the cutting edge ground with the rest unworked. Furthermore, these finds have so far remained isolated examples of early edge ground stone axes, without any evidence of further refinement, or spread, over wider areas.

In contrast, the polished hafted axe of the Fertile Crescent spread very rapidly around neighbouring areas followed by worldwide spread, including the Americas. At first glance it may appear that the polished stone axes of pre-Columbian America (*c.* 1500 BC), are examples of independent parallel invention or development, involving both, the polished stone axe as well as the Solutrean-type laurel-leaf point. Yet it is exactly the identity of their refinement with their African counterparts, and particularly their late appearance in America in which both appeared, which throws doubt on an independent American origin.

One may therefore suggest that when correctly assembled and classified in sequence, the history of stone tool-making the world over, illustrated and described in books and periodicals and exhibited in collections and museums throughout the continents, can be fitted into a framework comparable to the anatomical transformations in the animal kingdom. In outline the process involved allows a comparison with the Darwinian thesis of 'descent with modification', leading within the realm of stone tools to new genera, species, and varieties; with some species of stone tools ending in blind alleys. Darwin wrote (*Origin of Species*, 1859): 'The similar framework of bones in the hand of man, wing of bat, fin of porpoise, and leg of the horse – the same number of vertebra forming the neck of the giraffe and the elephant, and innumerable other such facts, at once explain themselves on the theory of descent with slow and successive modifications.' While the undertaking of such a systematic classification of stone tool developments would require a Darwinian patience and perseverance, it is important to point out that while animal kingdom evolutionary developments are biological, being based on genetical inheritance, no biological evolutionary, nor any cultural evolutionary process, within the kingdom of stone tools, is suggested here.

The situation is strikingly underlined by the proposition tentatively advanced in this paper that stone tool making throughout the millions of years of the stone age was the preserve of only a minority of culturally developing hominids. The majority of hominids although biologically identical with their stone tool making contemporaries, took little or no interest in stone tool making, nor in most other cultural developments. It can therefore not be asserted that a minority of stone tool makers were part and parcel of an evolutionary cultural process, while the majority (the none stone tool makers) were not. Finally, as a general rule, the archaeological record of stone tools conveys the impression that the majority of identical parallel developments in stone tool technology throughout the world resulted from continuing developments, which in each case ought to be traceable to a common source. Independent parallel emergencies of identical stone tool types, if they can be proved at all, are rare exceptions.

7

A Cultural Assessment of Homo Sapiens

Following a re-classification of fossils formerly listed under *Homo erectus*, latest estimates put the emergence of *Homo sapiens* at between 400,000 and 200,000 years BP, suggesting a safe medium of 300,000 years BP. They include finds at Verteszollos, Hungary; Swanscombe, England, and Steinheim, Germany. The Verteszollos fossils have been renamed *Homo sapiens paleohungaricus*, with an estimated age-tag of 200,000 years and a mean brain size of *c.* 1,400 cc. This compares with the contemporary *Homo sapiens* brain of 1,350 cc. (Bray/Trump, 1970; J. Campbell, 1983). Also of interest is the Swanscombe fossil, with an estimate age range of between 300,000 and 200,000 years, and a female brain size given as 1,325 cc. (von Bonin, 1963), which indicates a mean between male and female of 1,400 cc., this being well above present brain averages.

As to the achievements of culturally developing *Homo sapiens*, in terms of stone-tool development, to the best of our knowledge during his first 150,000 years (between 300,000 and 150.000 BP), they hardly exceeded the cultural achievements of their evolutionary predecessors *Homo erectus*. As Bray/Trump have pointed out (1970), the tools associated with *Homo sapiens paleohungaricus* (age *c*. 200,000 years), include small choppers made from pebbles and various flake tools. One can observe that both resemble the tools found at Bed II at Olduvai in E. Africa, which are a million years older. In Europe, *hungaricus* shows affinities with Clactonian tool types. In comparison, the earlier Swanscombe fossil was found with Middle Acheulian hand-axes, and at a lower level with Clactonian-type tools. Thus for *c*. 150,000 years or more, up to the advent of the Mousterian culture of the Neanders (at between *c*. 150,000 and 30,000 BP, *Homo sapiens* stagnated at the Acheulian level of culture already acquired by preceding *Homo erectus* a million years earlier. This in spite of the fact that at this distant age, *Homo sapiens* had already reached the average brain size of present humans. A cultural advance beyond the Acheulian was only achieved when the Neanders (now considered a sub-species or variety of *sapiens*) introduced the Mousterian culture.

In an assessment of the latter's physique D. Johanson writes (1982), 'I consider Neanderthal con-specific with *sapiens*. One hears talk about putting him in a business suit and turning him loose in the subway. It is true one could do it, and he would never be noticed.' What is more, according to W. W. Howell and E. Trinkaus (1980), their average brain size was 1,600 cc., well above that of modern man.

Yet during their *c*. 120,000 years of estimated existence (from 150,000 to 30,000 BP), the Neanders could not pride themselves of any dramatic cultural advance comparable with later Neolithic achievements. Some insight into their cultural limitations has been provided by Clark/Piggott (1979), who point to their apparent lack of concern with personal decorations, or art, and the total absence of perforated animal teeth, which later became the commonest of all ornaments worn by prehistoric hunters. Although the Neanders lived frequently in caves, the total absence of any cave art is another remarkable omission.

The Upper Palaeolithics who followed the Neanders biologically and culturally, were a younger version of *Homo sapiens*, preceding contemporary humans. Though producing a wealth of sophisticated artefacts and a greatly admired cave art, they were nevertheless culturally below their Neolithic successors. Thus, although they occupied large parts of the Old World for *c.* 30,000 years (between *c.* 40,000 to 10,000 BC) and their average brain size assessed by some authorities at 1,600 cc. (Thomas 1969 see under M. Day), they never achieved the late Neolithic and post-Neolithic feats of later *Homo sapiens* – culminating in food production and civilisation, all this occurring during a period of hardly 7,000 years, between *c.* 10,000 to 3,000 BC.

In considering the specific achievements of culturally developing *Homo sapiens*, two factors need stressing. One, that their presence covers c. 300,000 years, and two, that throughout this period average brain sizes remained fairly constant, except during Neander and Upper Palaeolitic times when they showed a conspicuous elevation, which subsequently reverted back to previous averages. Furthermore, no conclusive evidence has so far been produced to show that throughout this latter period (from *c.* 10,000 BC onwards to the present) the functional or intellectual (genetically related) value of the *sapiens*'s brain increased.

The *c.* 300,000 years of *Homo sapiens*'s life-span preceding food production, can perhaps be divided into *c.* 15,000 generations of 20 years each, with each generation assessed at 10 million people (see J. Campbell 1983). Accepting this figure we can arrive at a total world population of *c.* 7.5 billion having been present over this period of 300,000 years. In accordance with previous observations only a small part of this huge population can have taken part in cultural developments, while the rest persisted on the level of culturally under-developed, non-stone tool making hunter-gatherers, though, as a rule, using unworked stones.

The Meso-Neolithic period (*c.* 10,000 to 3,500 BC), which towards its end harboured the first stirrings of civilisation, spans *c.* 6,500 years, or 325 generations. By adhering to a world population rota of 10 million per generation, we obtain a total *Homo sapiens*, population of 3,250 millions for this latter period, lasting *c.* 6,500 years. A scrutiny of agricultural beginnings world wide, shows that only a small minority of these *Homo sapiens* could have initially engaged in food production, while the remaining billions remained food gatherers and hunters, some of them up to recent times.

Finally, the last developmental period to be reviewed is a span of mere 200 years (not more than 10 generations) between *c.* 3300 and *c.* 3,100 BC. marking the almost simultaneous unfolding of the ancient Sumerian and ancient Egyptian civilisations, both arising from an agricultural base. This sudden cultural explosion, was restricted to two relatively small riverine areas, a thousand miles apart. Indications are that everything that came after, was largely a blooming of the seeds that these two civilisations had planted earlier on, with one of them being the catalyst of the other.

Inquiring into the origins of food production, we have to ask (as in the case of stone tools), whether agriculture arose independently in many parts of the world or whether its beginnings are traceable to common sources of origin. J. Gowlett, who inclines towards the acceptance of many independent origins of agriculture, nevertheless expresses surprise at the almost simultaneous worldwide emergence of agriculture. He writes (1984): 'There is not much evidence of domestication before 10,000 years ago, but by 7,000 years ago cultivated crops and domesticated animals began to appear over large areas of the world, including America. All this did not come overnight, but in relation to two million years of hunting and gathering, the greatest ever alteration of economy came amazingly sudden.'

Considering the special case of cereal cultivation, it involves the selection of the right seeds from hundreds of wild grasses, and the growing of them in bulk, an altogether hazardous enterprise needing much experimentation. Once successful around 7.000 BC on the fringes of the Fertile Crescent of S. W. Asia, cereal cultivation, soon spread to almost everywhere. Yet throughout the entire period preceding 7,000 BC (or at most 10,000 BC) there was no known cereal cultivation anywhere in the world. In the Far East early cultivation of rice in China dates back to *c.* 5,000 BC while earlier claims for Thailand (*c.* 6,800 BC – Spirit Cave), have remained dubious. In Africa earliest cereal cultivation is reported from the Fayum Oasis dating back to 5,200 BC (Trigger, 1983). In the Americas, earliest maize cultivation dates back to *c.* 5,000 BC in Mexico (Tehuacan Valley). In Europe earliest traces of cultivation have been found in Cyprus, dated at *c.* 6,100 BC and in Thessaly (Greece), dated at *c.* 5,500 BC. The most significant omission from this list is the continent of Australia, where cereal growing had to await modern European colonisation.

What all these dates show is that the world's early cereal cultivation fell within a time range of hardly 2,000 years (i.e. from *c.* 7,000 – 5,000 BC). Yet preceding this, *Homo sapiens*, throughout his *c.* 300,000 years' history (minus 10,000 years' agricultural history), never succeeded in cultivating cereals anywhere (although possibly there were abortive attempts). And yet during this huge expanse of time there must have occurred numerous occasions when climatic and soil conditions, and the presence of suitable grass seeds must have favoured cereal cultivation. For example, J. Gowlett (1984) points out that just over 120,000 years ago, for about 10,000 years, the climate in most of the world was very similar to that of today, in land, sea, and vegetation. Yet there was no sudden development towards agriculture and civilisation.

In view of this the following proposition stands to reason. Presuming that during the 300,000 years of *Homo sapiens*'s existence (preceding agriculture) there was, arbitrarily, at least one favourable occasion every hundred years to grow cereals, there could have been at least 3,000 such occasions. Yet, as far as we know, not one such opportunity was utilised. The assumption therefore, that the world's cereal cultivation started up almost simultaneously and independently in many widely separated areas of the world within a narrow time span of 2,000 years is inconceivable.

Many years ago Gordon Childe highlighted this situation when he wrote (1966): 'It must not be imagined that at a given moment in the world's history a trumpet was blown in heaven, and every hunter from China to Peru thereupon flung away his weapons and traps and started planting wheat or rice breeding pigs, sheep and turkeys.' George Carter observed (1973b:4):

All men were seed-gatherers for millions of years and were subject to environmental and population stress and almost all men had the metate for at least tens of millennia. Yet at most, in only a few spots, and in terms of the time scales we are dealing with, at about the same time, men suddenly undertook the domestication of plants. Why, if the need, opportunity and focus of interest, had existed for millions of years, (and the human ability of *Homo sapiens* with brain capacities approaching ours, had existed for about half a million years), should this rush of domestication have burst out all over the globe like an epidemic of measles? Was it indeed like measles, a communicable disease?'

In contrast, J. Gowlett has tried to explain why many archaeologists accept independent agricultural origins. He points out that modern humans before taking to agriculture gradually had accumulated similar levels of cultural experience, had encountered similar problems and had affected similar adaptations. But has this really been the case? The archaeological record shows otherwise.

Before agriculture took hold the world was exclusively populated by hunter-gatherers, the majority of whom had never advanced to the alleged level of cultural experience (mentioned by Gowlett), which ought to have led them almost automatically into agricultural pursuits; and this, although throughout hundreds of thousands of years they had encountered similar foraging problems and lived in the same environment as their culturally more advanced contemporaries, while biologically they stood all on the same *Homo sapiens* level.

Furthermore, the theory claiming an almost automatic progress towards agriculture, is contradicted by the Australian experience. According to Gowlett himself (1984), one group of South-East Asian hominids crossed into Australia (via New Guinea) as early as 70,000 to 60,000 years ago, while another followed *c.* 20,000 years later. Neither group ever practised agriculture. Yet the prevailing conditions in Australia ought to have favoured such a development, as along with other cultural elements, Gowlett reports the discovery of grinding stones dated *c.* 15,000 BP, commenting that their presence suggests the exploitation of seeds and their grinding.

In both the Old World and the New the presence of grindstones and querns, indicating the collection of wild growing grains and their grinding into flour, (dating in the Old World well back over 200,000 years (Carter: 1973a)), has long been considered as a prelude to the deliberate cultivation of cereals (i.e. agriculture). In Australia such grain collection and grinding is also evidenced but led to no further development. When grain cultivation was eventually introduced into Australia in the wake of European colonisation about two hundred years ago, the continent became one of the world's leading grain producers.

The conclusion follows, that when all the evidence for and against the independent parallel emergence of agricultural beginnings (especially for grain cultivation) is considered, the indications for a single origin followed by world wide spread from this source far outweigh anything advanced in favour of multiple independent origins.

George Carter, in 'A Hypothesis Suggesting a Single Origin of Agriculture', came to similar conclusions.* Carter argued that potential domestic plants are ubiquitous, but only a few plants were originally chosen and the New World plants mimic the Old World domestics. The whole thing happened worldwide at the same time (give or take a couple of thousand years), Carter remarks that the idea of a possible single origin of agriculture goes back to G. Eckholm who muted it 25 years earlier.

* Published 1977 in *Origin of Agriculture*, by Charles A. Reed (Aldane, Chicago pp. 83–133).

8

The Case of Early Civilisations

The term 'civilisation' is subject to differing interpretations. Prominent pre-historian, Professor Fred Wendorf of Southern Methodist University Dallas, Texas, among others, would rather avoid the term altogether, preferring to speak instead of 'complex societies'.

Here we retain the term, though limit it specifically to those earliest, as well as later civilisations, which can be described as 'urban literate'. Their prototypes are the civilisations of Egypt and Sumer, which emerged almost simultaneously between *c.* 3,300 and *c.* 3,100 BC; both contain cultural elements of great likeness, in their details of material culture, and their ideational components. Other early civilisations differing only in detail, but not in substance, emerged subsequently in India, China, Mexico (including Mesoamerica) and Peru.

Spectacular achievements in the Old World include monumental temple structures, pyramids and palaces and the emergence of large cities. They share the use of writing (though not evident universally) and the increased use of gold, precious or semi-precious stones and metals.

As to the special case of Sumer and Egypt, both emerged within a span of a mere 200 years. In this short period, both civilisations appear to have displayed many of the basic traits which led to their later splendour. Their formative stages dating back to perhaps 4,000 BC and before, are still only vaguely discernible, and, although present evidence favours Sumer as having been the earlier civilisation, there are indications that Egyptian priority cannot be ruled out.

The essential accomplishments which separate these two early civilisations from preceding less complex societies, include the use of writing, the first recorded emergence of kings and gods (secular and divine), organised religion and its priesthood, large cities with thousands of inhabitants, monumental edifices, including temples, palaces and pyramids, and in Egypt, the early use of a calendar. It can be presumed that the remains of many ancient Egyptian cities are still buried under the Nile Valley's alluvium.

However, the appearance of many of these traits was not uniform in the two civilisations, neither chronologically nor in the sharing of common cultural elements. Furthermore, in the course of time each civilisation developed its own individual character. Yet it is unlikely that such a cluster of distinct cultural elements held in common, could have arisen independently of each other, either singly or in combination, during the brief time these two civilisations reached prominence. Geographically, they lay about a thousand miles apart and it is notable that a depiction of Sumerian-type boats in pre-Dynastic Egypt bears witness that inter-communication between the two existed since early times. J. Gowlett (1984) has observed that writing appeared at about the same time (between 3,100 and 3,000 BC), both in Mesopotamia and ancient Egypt. Since it appeared in Egypt, seemingly without any sign of preceding development, some authorities have suggested that the concept may have been imported from Mesopotamia.

Only a few hundred years later, similar civilisation arose in India and China, and like their predecessors in Sumer and Egypt, they were situated in big river valleys, i.e., Indus and Hoangho. Like their Sumerian and Egyptian antecedents, both India and China had early scripts, though the Indian script has so far defied decipherment. J. Gowlett says that the fact that writing appears several hundred years later in India and China than in Mesopotamia suggests that the inspiration for it came to them from Sumer.

All these four Old World civilisations (Sumer, Egypt, India and China) were distinguished by large urban settlements, some of them still awaiting excavation. The architecture of pyramidal structures in China resembles that of the step-pyramids of Sumerian Ziggurats which again may go back to the ancient Egyptian step pyramid of Pharaoh Zoser of the Third Dynasty. More remarkable is that similar step-pyramids are a prominent feature of Pre-Columbian American civilisations. In fact, in all early civilisations with the exception (so far) of India, we find an early presence of gods, kings and organised religion, administered by a priestly class. In each case (again in India they are noted later) we find whole city states or nations ruled over by powerful kings, some of them worshipped as gods in their lifetime. These often formed dynasties, perpetuated by hereditary succession, the most prominent examples being the Pharaohs of Egypt ('Sons of the Sun' from the V Dynasty onwards), the Sons of Heaven in China, and the Inca ruler in Peru, claiming to be a 'Son of the Sun', and the Mikados of Japan, reputed to have descended from the sun goddess Amaterasu; they were worshipped as 'Living Gods' up into modern times.

It is now generally conceded that there were many links between the Sumerian, Indian and Chinese civilisations, just as there were striking similarities between Sumer and Egypt. This is hardly surprising as for thousands of years before their emergence we find evidence of the movements of people and of trade covering enormous distances. For example, identical Venus figurines of the lower Palaeolithic Gravettian culture (between 30,000 and 25,000 BC) have been identified at distances *c*. 3,200 km apart – from the foothills of the Pyrenees in Spain to the area of the lower Don in S. Russia. Cave art of a Magdalenian pattern (well before 10,000 BC), comparable in style, has been evidenced 4,000 km apart, in the Dordogne in France and in the S. Urals (Clarke/Piggott, 1970). More recent trading links have been mentioned by J. Gowlett (1984), for the period between *c*. 7,000 and 5,000 BC in such goods as seashells from the Indian Ocean to Mergarah (an outpost of the Indus Civilisation) *c*. 800 km to the north, and of lapis lazuli from

Badakshan in Central Asia to both, Mergarah 1,000 km to the south and via Afghanistan to Mesopotamia, a distance of 2,500 km. A lively trade in obsidian, from the Lake Van area in Anatolia to both Southern Mesopotamia and to Palestine is also recorded.

In view of these facts, the postulation by some adherents of the theory of independent cultural development, namely that early civilisations, despite sharing many identical features, were in each case the isolated creations of their local inhabitants, appears no longer valid.

The manner in which this process of independent isolated parallel development is presumed to have occurred has been elucidated by Glyn Daniel, a leading exponent. He reasoned (1971) that 'seven societies in seven different ways trod the paths that led to civilisation', basing his proposition on ideas earlier advanced by Kroeber and Caldwell. Kroeber had written (1940): 'We must consider that civilisation is an inevitable response to laws governing the growth of culture and controlling the man-culture relationship.' And Caldwell (1966): 'Perhaps there is only a finite number of social and historical processes behind the events of history.' Daniel concluded (1971): 'I believe that an interpretation of the origin of civilisation in terms of multi-linear evolution [i.e., conforming with the ideas of Kroeber and Caldwell, G.K.], is in accordance with the archaeological facts as known to us.'

In the spirit of modern scientific inquiry, this deterministic metaphysical approach to human history does not conform with archaeological facts, as Daniel maintains. Under the conditions outlined by Daniel and Gowlett, two of the most advanced pre-civilised societies in the Near East, Jericho in Palestine, c. 8,800 to c. 6,000 BC, and Catal Huyuk in Anatolia c. 6,750 BC and after, should have developed into fully fledged civilisations. Both went through the stages of a food-collecting economy, leading eventually to mixed farming and pottery making, with Catal Huyuk even developing irrigative agriculture. Both these sites developed into large settlements, with many examples of artistic expressions and with Catal Huyuk eventually occupying an area four times the size of Jericho.

However, neither of them erected monumental temples or palaces. At Catal Huyuk rooms of modest size have been excavated (decorated with bulls' heads), which have been described as temples. There is no evidence of any hierarchical structure of government headed by kings or queens, no evidence of gods and organised religion. Also at Catal Huyuk there are several statues of big females, sitting on thrones, who have been called Mother, or Fertility Goddesses. Neither of the two societies used writing or a calendar. In fact, Jericho, on its own account, never reached a higher level of culture akin to civilisation, while Catal Huyuk petered out of history and left nothing but ruins, recalling a Neolithic type of advanced farming society. Mellaart, discoverer and excavator of Catal Huyuk wrote (1965:75): 'The neolithic civilisation [read pre-civilisation, G.K.] revealed at Catal Huyuk, shone like a super-nova ... it burned itself out and left no permanent mark on the cultural development of Anatolia after 5,000 BC.'

In contrast with the favourable conditions prevailing at Jericho and Catal Huyuk, Sumerian civilisation established itself on what was originally a swampy wasteland without any trace of preceding cultural development, going much beyond the level of reed-hut fishing villages, presumed to have resembled those of contemporary Marsh Arabs. This in an area not only devoid of wood, but also of stone and metals. It was in such a precarious environment that Sumerian civilisation began, with Eridu (reputed to be the first city in the world) apparently establishing a tenuous foothold on an elevated sand dune adjoining the Persian Gulf. Seton Lloyd and F. Safar (1981), point out that 'the first human settlement was located at a site now called Tell Abu Sharein. It was a high dune of wind drifted sand, possibly forming an island in a wide area of marshland once a tidal lake at the head of the Persian Gulf.'

The physical pre-conditions of the area where the first American civilisation, that of the Olmecs, became established, c. 1,200 BC (M. Coe, 1968), appear to have been equally unpromising. Coe writes that 'the abruptness of its appearance at the hot coastal plain of Southern Vera Cruz has no convincing explanation at the present moment.' A much later writer, Stuart Fiedel (1987), comments that 'Much work remains to be done before we can fully understand why Olmec civilisation should have arisen so precociously in a seemingly inhospitable environment.' And Jacques Soustelle, writing on the origin of the Olmecs (1985), points out that 'There is no evidence of "formative" evolution, a gradual maturation over several centuries. This indeed constitutes the very head of the Olmec mystery … the astonishing spectacle of a civilisation that gives the impression of suddenly springing up in all its originality from an undifferentiated background of peasant culture.' Soustelle concludes: 'We are naturally left to ponder the question whether this leap was not due to exterior influences?'

In this respect the finding of R. A. Jairazbhoy, who has traced Olmec origins to ancient Egyptian parallels, are most revealing. Jairazbhoy (1974) shows the presence of ancient Egyptian traits in Olmec Mexico dating from the time of Rameses III onwards. These include many identical replications of Egyptian gods, some Egyptian hieroglyphs, and a great deal more.

These few instances of factual evidence cast serious doubt on the validity of the multi-linear evolutionary approach, which pretends to postulate the independent, isolated origins of the early civilisations in both the Old World and the New.

What then is the alternative?

I believe Chapter 6 ('An Anatomy of Stone Tools') presents a fairly convincing case for a continuing development of stone-tool technology, showing the dissemination of specific tool types from common sources of origin to widely separated parts of the world. We can propose that a similar process of spread from common origins predominates and explains the presence of other cultural parallels the world over, beginning with the Old Stone Age and ending with modern civilisation. Extensive supportive evidence for this is being submitted in Part Three.

Bibliography for Part One

Barnett, A., 1964, *The Human Species*, Collins

Bailey, Geoff, 1983, *Hunter-Gatherer Economy in Prehistory*, Cambridge University Press

Brace, C. and Montague, A., 1968, *Man's Evolution*, Collier

Bray, W. and Trump, D., 1970, *The Penguin Dictionary of Archaeology*

Bridewood, and Willey, 1962, *Courses Towards Urban Life*, Edinburgh University Press

Bonin, G. von, 1963, *The Evolution of the Human Brain*, University of Chicago Press

Caldwell, J. R., 1966, *New Roads To Yesterday*, Thames & Hudson

Clark, D., 1968, In *Symposium*: 'Man The Hunter'.

Campbell, B., 1974, *Human Evolution*, Aldine Publishing Co., Chicago, USA

Campbell, J., 1983, *The Way of the Animal Powers*, Harper & Row, San Francisco

Childe, E. G. 1966, *Man Makes Himself*, Fontana Library

Cippola, C. M., 1965, *The Economic History of World Populations*, Penguin Books

Clark, G. and Piggott, St, 1970, *Prehistoric Societies*, Penguin Books

Clarke, R. J., *see under* Grusser O. J., 1985

Coon, C. S., *The History of Man*, Penguin Books

Daniel, F. Glyn, 1971, *The First Civilizations*, Penguin Books
Dart, R. A., 1957, *The Osteodontokeratic Culture of Australopithecus*, Transvaal Museum Mem. 10, South Africa
Darwin, Charles, 1859, *Origin of Species*, I, John Murray
— 1866, *Origin of Species*, I, John Murray
— 1958, *Origin of Species*, World Classics, Oxford, University Press
— 1859, *The Descent of Man*, John Murray
Day, Michael H., 1977, *Guide to Fossil Man*, Cassell Co.
Day, Michael H., 1973, *Human Evolution*, Society for the Study of Human Biology, vol. 11, Taylor and Francis
Day, M. H., Leakey, M. D., and Olson, T. R., 1980, *On the Status of Australopithecus Africanus*, Science, pp. 1102–3
Encyclopaedia Britannica, 1961 edition
Encyclopaedia of Anthropology, 1976, P. Harper & Row, New York, USA
Falk, Dean, 1980, *Hominid Brain Evolution*, Alan R. Liss, New York, USA
Fiedel, S. T. 1987, *Pre-history of the Americas*, Cambridge University Press
Flood, H. 1983, *Archaeology of the Dream Time*, Collins
Forde, Daryl, C., 1968, *Habitat Economy and Society*, Methuen
Franzen, J. L., *See under* Grusser, O. J., 1985
Goodall, G., van Lawick, 1964, *Tool Using of Free Living Chimpanzees*, Nature, London 201, 1264–6
Gowlett, J., 1984, *Ascent to Civilization*, Collins
Grusser, O. J. and Weiss L. R., 1985, *Hominid Evolution, Past, Present and Future*, Proceedings Taung Diamond Jubilee. A. R. Liss, New York
Guilmet, G. M., in *Man (1977, vol. 12, no. 1)*
Harris, M., 1968, The Rise of Anthropology, Routledge and Kegan Paul
Harrison, A. J., 1969, *Man the Peculiar Animal*, Penguin Books
Holloway, R. L., 1975, *The Role of Human Social Behaviour in the Evolution of the Brain*, 43rd J. Arthur Lectures, 1973 – 1978, *The Relevance of Endocasts for Studying Primate Brain Evolution*, Plenum Publishers, New York pp. 181–200
— 1979, *Brain Size, Allometry, and Reorganization towards a Synthesis*. Academic Press, New York, pp. 59–88

— 1984, 'The Poor Brain of Home Neanderthalensis' in *'Ancestors' – The Hard Evidence*, Allan R. Liss, New York

Howells, W. W., 1967, *Mankind in the Making*, Penguin Books

Howells, W. W., and Trinkaus, 1980, *The Scientific American*, pp. 94–105

Huxley, J. S., 1957, *Evolution in Action*, Mentor Books, USA

— 1963, *Evolution: The Modern Synthesis*, Allan & Unwin

Isaac, G. L., 1978, 'Archaeological Evidence of the Activities of Early African Hominids', in *Early African Hominids* (ed. C. J. Jolly, Duckworth)

Jairazbhoy, R. A. 1974, *Ancient Egyptians and Chinese in America*

Jerison, H. J., 1970, *Gross Brain Indices and the Analysis of Fossil Endocasts*, The Private Brain, 1255–44, New York

— 1973, *Evolution of the Brain and Intelligence*, Academy Press, USA

Johanson, D. C., and Edey M. A., 1982, *Lucy*, Granada Publishing Ltd

Koestler, A., 1967, *The Ghost in the Machine*, Hutchinson

Kraus, G., 1973, *Homo Sapiens in Decline*, New Diffusionist Press

— 1977, *Man in Decline*, St Martin's Press, New York

Kroeber, A. L., *see under* Daniel G., 1971

Leakey, M. D., 1971, *Olduvai Gorge* vol. 3, 'Excavations in Beds I and II,' Cambridge University Press

Leakey, R. E., and Levin, R., 1981, *The Making; of Mankind*, Michael Joseph

Liam de Paor, 1971, *Archaeology*, Pelican Original

Lloyd, Seton, 1978, *Archaeology of Mesopotamia*, Thames & Hudson

Lloyd, Seton and Safar, F., 1981, *Eridu*, Baghdad (final official report) McHenry, H. M., 1982, 'The Pattern of Human Evolution', *American Review of Anthropology* II, 151–173, USA

Mellaart, J., 1967, *The Earliest Settlements in Western Asia*, Cambridge University Press

Parker, S. T., and Gibson, K. R., 1979, *A Development Model for the Evolution of Language and Intelligence in Early Hominids*. The Behavioral and Brain Sciences 2 (367–381)

Passingham, R. E., 1975, *The Brain and Intelligence*, 2: 499–508

Pilbeam, D., 1960, The Evolution of Man, Thames & Hudson

Radinsky, L. B., 1979, *The Fossil Record of Primate Brain Evolution*, 49. James Arthur Lecture, American Museum of Natural History

Rench, B., 1959, *Evolution Above the Species Level*, Methuen

Rhodes, R. T. H., 1962, *The Evolution of Life*, Penguin Books

Russel, Sir, E. John, 1966, *The World of the Soil*, Penguin Books

Shrire, Carmen, 1983, *Past and Present in Hunter-Gatherer Studies*, Academic Press N.Y./London – Symposium Bad Homburg, Germany

Simpson, G. G., *The Major Features of Evolution, Columbia University Press, New York*

Soustelle, J., 1985 (1979), The Olmecs, Doubleday, New York

Tobias, P. V., 1970, *New Endocranial Volumes of Australopithecines*, Nature 227: 199-200

— 1971, *The Brain in Hominid Evolution*, Columbia University Press, New York

— 1975, *Brain Evolution in the Hominidae*, Mouton Publisher, The Hague, Holland

— 1982, *The Antiquity of Man*, Alan R. Liss, New York

— 1983a, *Recent Advances in the Evolution of the Hominids*, Pontifical Academy of Science – Scipta Varia, vol. 50, pp. 85–140

— 1983b, *Late Cainozoic Paleoclimates of the Southern Hemisphere*, Symposium, South African Society, Rotterdam/Boston

— 1983c, *South African Hominids and the Evolution of Man, 'Terra' (pp. 11–15)*

– 1985a, *Punctuational and Phyletic Evolution in the Hominids*, 'Verba', p. 131–41, Transvaal Museum, Pretoria, S. Africa

— 1987, 'The Brain of Homo Habilis', *Journal of Human Evolution*

Trigger et al., 1983) *Ancient Egypt*, Cambridge University Press

Turnbull, Colin M., 1966, *Wayward Servants*, Eyre and Spottis-wood

— 1976, *Man in Africa*, Anchor Press, Doubleday, New York

— 1983, *The Mbuti Pygmies, Change and Adaptation*, Holt, Reinhard and Winston, New York

White, Carmel, 1967, In *Antiquity* 1967, vol 41, 49–52
Willey, G., *see* Bridewood and Willey, 1962
Wolberg, D. L., 1970. 'Comment on Dart in *Current Anthropology*
    II, 23–37

Part Two

A Scrutiny of Cultural Theories

9

Multilinear Cultural Evolution

To reiterate, nowadays practically all social anthropologists and ethnologists reject the Lamarckian-linked theory of unilinear cultural evolution as obsolete. This includes leading multi-linealists such as A. L. Kroeber (who had earlier revoked it), as well as J. R. Caldwell, J. H. Steward, and Glyn Daniel. But while they have abandoned unilinealism in favour of multilinealism, it can be shown that while rejecting the former theory in its abstract, they still follow its basic tenets in practice.

Glyn Daniel, in *The First Civilizations* (1971), deals competently with the archaeological background preceding the emergence of early civilisations. But in his conclusions relating to the actual origin of these civilisations he vacillates. In his previous work, *The Idea of Prehistory* (1964:91), he defined unilinear cultural evolution (which he rejects) as 'a form of super-organic or cultural or social evolution'. Seven years later in *The First Civilizations*, he described multilinear evolution in practically the same terms, as 'a supra-organic, a cultural evolution in man's development'; calling it: 'the basic principle of the culture-process which leads to civilisation' (1971:176-177). A closer examination reveals that both definitions are identical, and since in the same paragraph, Glyn Daniel attributes a like trend of thought to other multilinealist colleagues, such as Kroeber and Caldwell, we can see that all these cultural theorists are unable to clearly distinguish between what is unilinear and what is multilinear in cultural evolution. That both are in fact identical is confirmed by Leslie White, the doyen of cultural evolutionism. He wrote (1959:30/31): 'Evolutionist interpretations of culture will be both unilinear and multilinear. One type of interpretation is as valid as the other, each implies the other.'

Furthermore, according to dictionary definition, both terms 'super-organic' and 'supra-organic' are conceptions which reach outside, or beyond the realms of any earthly, organic, or biological existence. Chambers *Twentieth English Dictionary* (1960 edition) defines the term 'super-organic' (which is the same as supra-organic – G.K.) as: 'above or beyond the organic, psychical, pertaining to higher organization'. To somehow mitigate this metaphysical aspect, multilinear cultural evolutionism as explained by Daniel, holds that unilinealism (the discarded thesis) is universally applicable, while multilinealism (its acceptable replacement) has a restricted, selective application. J. H. Steward's definition of 'multilineal cultural evolution' (1955:19), reads as follows: 'Multilineal evolution is not interested in particular cultures, but is interested in finding local variations and diversity, troublesome facts which force the frame of reference from the particular to the general; it deals only with those limited parallels of form, function and sequence which have empirical validity.' Altogether a truly Delphic phraseology. In effect, the above demonstrates that both unilinear and multilinear cultural evolution are subject to the same super-organic, metaphysical guiding principles, placing them firmly outside any biological evolutionary conception.

This obviously determinist interpretation of multilinealism is endorsed by its originators. Kroeber (1940) maintained that: 'We must consider that *civilization is an inevitable response to laws governing the growth of culture and controlling the man-culture relationship*'; and Caldwell: 'Perhaps there is only one finite number of social and historical processes *behind the events of history*'; while Daniel comments (1971:176): 'I think Caldwell and Kroeber are right … We should now think in terms of multilineal evolution, *leading inevitably*, as Kroeber said, for some ... societies with geographical and ecological and cultural possibilities to synoecism, one of the finite number of social and historical *processes behind the events of history*.' Daniel concludes: 'I believe that an interpretation of the origins of civilisation in terms of multilineal evolution is in accordance with the archaeological facts as known to us.' Again, in discussing the origins of American civilisation, Glyn Daniel (1964:105), asserts that: 'it was in fact a tale of independent cultural evolution.'

The above italicised phrases express the true nature of multilinear thinking, whereby multilinealist cultural evolution is revealed as a concept guided throughout by determinist super-organic forces, which, in substance places it in the sphere of the paranormal i.e. metaphysics.

It is also notable that none of the preceding expositions on multilinear cultural evolution mention any biological linkage in terms of modern genetics, a trend apparent in the work of J. H. Steward, a leading theorist on the subject. While discussing at length cultural evolution vis-à-vis biological evolution, he concludes (1953:11/14), that the former (i.e. cultural evolution) has no biological association. This is in accord with Daniel's previously mentioned definition of a 'super', or 'supra-organic' process. Steward further says that: 'The mythology of evolution contains two vitally important assumptions. First, it postulates that genuine parallels of form and function developed in historically independent sequences or cultural traditions. Second, it explains these parallels by the *independent operation of identical causality*.'

Again, the italicised words point to a process linked to Lamarckism, based on the fact that culture is created by humans and they could only produce parallel cultural traits independently in response to identical conditions of life (i.e., identical causality in a Lamarckian sense); this in turn depends on hereditary predispositions moulded by 'identical causality'. In other words, man's parallel cultural expressions are said to be shaped by acquired characteristics which are inheritable, this being an example of the equal working of the human mind in response to like circumstances. In short, we have here an endorsement of Bastian's principle of the 'psychic unity' attributable to all mankind. In its essence it implies a biological process based on Lamarckian inheritance.

Steward elaborates this Lamarckian reasoning as follows:

> Cultural evolution, then, may be defined broadly as a quest for cultural regularities or laws; but there are three distinctive ways in which evolutionary data may be handled. First, unilinear evolution, the classical nineteenth-century formulation, dealt with particular cultures, placing them in stages of universal sequence. Second, universal evolution – a rather arbitrary label to designate the modern revamping of unilinear evolution – is concerned with culture rather than with cultures. Third, *multilinear evolution*, a somewhat less ambitious approach than the other two, is *like unilinear evolution* [my italics] in dealing with developmental sequences, but it is distinctive in searching for parallels of limited occurrence instead of universals.

Thus multilinear evolution is presented here as a variation of unilinear evolution, a view endorsed by Lesley White (see above).

The passage from Steward cited above (1953:14) is followed by many pages of involuted reasoning attempting to substantiate his view, but in effect carrying very little conviction. I find it appropriate to include here some criticism by Colin Turnbull of the ethnological theories Steward put forward in his work *Theory of Culture Change* (Urbana, USA, 1955). Turnbull comments that they are 'unsupported assertions, derived from a superficial treatment of the subject, based on incorrect information' (1966:277).

In the light of such attempts the conception of multilinear cultural evolution presents a remarkable aberration in ethnological thought. It can perhaps be explained by its proponents conjuring a smokescreen against cultural diffusionist explanations, and in the process taking recourse to irrational arguments and verbally elusive pseudo-scientific jargon.

This may appear a rather harsh judgement, although it is far less outspoken than the comments of the great scholar Bertauld

Laufer. In a review of Lowie's 'Culture and Ethnology' (*American Anthropology* 1918, 20:87/8) Laufer wrote (and I repeat):

The theory of cultural evolution, to my mind, is the most inane, sterile and pernicious theory ever conceived in the history of science (a cheap toy for the amusement of big children) ... Culture cannot be forced into a strait-jacket of any theory whatever it may be, nor can it be reduced to chemical or mathematical formulae. All that the practical investigator can hope for, at least for the present, is to study each cultural phenomenon as exactly as possible in its geographical distribution, its historical development and its relation or association with kindred ideas.

10

The Structuralist Approach

Structuralism is basically an attempt by ethnological theorists, explicitly differing with diffusionist and evolutionist interpretations of social and cultural phenomena, to replace or supplement the former by structuralist solutions. They assert that common social phenomena like beliefs, institutions, customs, rituals, myths, totems and taboos, can be explained by the working of underlying 'deep' or 'hidden' structures such as collective dreams and unconscious wishes. Levi-Strauss, the leading structuralist, assumes with Freud (Leach, 1970:57) that 'a myth is a kind of collective dream, and that it is capable [by the analysis of its elementary structure – G.K.] of revealing hidden meanings. According to Freud, myth expresses unconscious wishes which are somehow inconsistent with conscious experience.' Leach (1970:105) further comments that 'it becomes increasingly difficult to understand just what Levi-Strauss means by "elementary structures".' Levi-Strauss also assumes with Freud, (Leach 1970:57) that 'the "incest taboo" is a kind of collective dream, and that it is the cornerstone of all human society.' French anthropologist Levi-Strauss (1968:3) has maintained that the structuralist approach to culture has no cultural evolutionist associations. An analysis of its basic concepts as given above suggests otherwise.

While structural analysis (originally used in linguistics),

presents a quite novel approach to cultural theory, it gained notoriety when applied and expanded upon by Levi-Strauss, who is perhaps better described as a structuralist philosopher. Some of his expressions as exemplified by Edmund Leach are 'products of ethnological fantasies straying into poetry'.

Leach, a prominent English theorist who is both an admirer and critic of Levi-Strauss, has praised the truly poetic range of associations which Levi-Strauss brings to bear in the course of his analyses (1970:118), but at the same time he subjects Levi-Strauss to harsh criticism. Leach (1970:20) on Levi-Strauss: 'Any evidence however dubious is acceptable so long as it fits with logically calculated expectations; but wherever the data run counter to the theory Levi-Strauss will either bypass the evidence or marshall the full resources of his powerful invective to have the heresy thrown out.'

Levi-Strauss adopts a more realistic approach when dealing with 'parallel art forms' under the heading 'The Split Representations in the art of Asia and America'. He writes (1968: 246/248):

> It is impossible not to be struck by the analogies presented by the Northwest Coast and ancient Chinese art. These analogies derive not so much from the external aspect of the objects as from the fundamental principles which an analysis of both arts yields.
>
> . . . Once these similarities have been noted, it is curious to observe that, for entirely different reasons, ancient Chinese and Northwest Coast art have been independently compared with Maori art in New Zealand. . .
>
> Do we rest, then, on the horns of a dilemma which condemns us to deny history or to remain blind to similarities so often confirmed?
>
> . . . We reserve, therefore, the right to compare American Indian art with that of China or New Zealand, even if it has been proved a thousand times over that the Maori could not have brought their weapons and ornaments to the Pacific Coast. Cultural contact doubtless constitutes the one hypothesis which most easily accounts for complex

similarities that chance cannot explain. But if historians maintain that contact is impossible, this does not prove that the similarities are illusory, but only that one must look elsewhere for the explanation. The fruitfulness of the diffusionist approach derives precisely from its systematic exploration of the possibilities of history. If history, when it is called upon unremittingly (and it must be called upon first), cannot yield an answer, then let us appeal to psychology, or the structural analysis of forms; let us ask ourselves if internal connections, whether of a psychological or a logical nature, will allow us to understand parallel recurrences whose frequency and cohesion cannot possibly be the result of chance.

This resource to psychology as a final arbiter to problems which Levi-Strauss feels cannot be resolve by diffusion, re-occurs in most of the other Levi-Straussian ethnological theories, whether they deal with kinship systems, myths, religion, or pure semantics.

Discussing Levi-Strauss's pre-occupation with kinship-systems, Leach observes (1970:95): 'Although there are thousands of different human languages, all kin term systems belong to one or other of about half a dozen "types", how should we explain this? Levi-Strauss does not follow Lewis Morgan at all closely but he assumes, as we might expect, that any particular system of kin terms is a syntagem of the "system" of all possible systems, which is in turn a precipitate of a universal human psychology.'

Such reliance on psychology is a return to Bastian's (Lamarck-based) 'psychic unity of all mankind', where it forms the basis of unilinear cultural evolutionism, a principle continued by Frazer, Freud. Jung and others and now endorsed by Levi-Strauss. The fact that the bulk of mankind, the primitive food-gatherers of all ages (over millions of years), are excluded from this assumedly universal state of affairs, renders it illusionary. A few brief examples of present survivals can substantiate this. Morris (1976:543) found many Hill Pandaram (India-based primitive food-gatherers) could not remember the names even of their grandparents, a condition which Turnbull had also noted among the Mbuti pygmies of the Congo. Leach (1970: 105) has pointed out that the majority of what are by some ethnologists usually considered to be surviving 'ultra primitives' (e.g. Congo's Mbuti pygmies and Kalahari Khoi-San-Bushmen) do not have kinship systems of unilinear descent (considered to be the simplest form of this genre – G.K.). Leach, in a private communication, informs me that he personally rejects the term 'ultra primitive', for describing culturally less complex societies or peoples. Finally, I repeat Turnbull's remarks on kinship amongst the Mbutis: 'With the best will in the world, it is not possible to apply the kind of analysis suggested by Levi-Strauss (1958), to the Mbuti (1977:238).'

A more likely, historical explanation for the origin and continuity of kinship and lineage systems has been proposed by C. E. Joel in *New Diffusionist* (1970, Nos. 1 and 3). Referring to Hocart, (*The Life-giving Myth*, Chapter XXII), Joel attributes them to the emergence of early kingship when royal family relationships began to play an important part, involving royal succession with its transfer of power and prestige and the inheritance of property. Joel (1970:87) points out that these and related factors arouse the strongest passions in every society. These originally restricted kinship relations became modified in the course of time as they devolved down to more primitive, semi-civilised, sedentary and pastoral client groups.

This assessment finds support in a detailed examination of kinship and lineage principles. In Macmillan (1983:183), it is pointed out that 'among the central components of kinship are grant and social paternity for the purpose of inheritance.' It is also stated that Radcliffe-Brown and Fortes in their kinship theories stress inheritance and succession (p.184). Goody suggests that descent theories use examples involving homogenous transmission of property after death (p.184)

As to lineage relations (Macmillan p. 209), it is stated that 'in principle, patrilineal systems depend upon the passing of status and property from father to legitimate son; and in matrilineal systems inheritance is traced from the uncle to nephew.' Thus the entire field of kinships and lineages is permeated by property, status and inheritance. Such conditions apply however solely to property-owning sedentary and pastoral societies which only emerged in the wake of food production. They are, in terms of *total* hominid history, a rather novel and recent phenomenon. The circumstances of their appearance will be treated in more detail in my next exposé.

In summary, the conditions which form the basis of kinship and lineage structures can certainly not apply to culturally ultra-primitive food-gathering groups, to whom land ownership, inheritance of property and private property itself (beyond modest personal possessions) means nothing.

Colin Turnbull (1966:238) writing about the social structure among the Mbuti of the Congo, has this to say: 'The band is the basic unit and its membership is influenced by economic rather than kinship considerations', and (1965) 'The Mbuti show no tendency to adopt or develop the more complicated forms of social organisation possessed by their neighbours (i.e. their Bantu masters). When they do adopt some customs of their masters, they do so for purely opportunistic reasons.'

In conclusion, 'structuralism' suffers from some of the same weaknesses as cultural evolutionary theories. It is unable to explain rationally the independent emergence of cultural parallels in a worldwide context, be they of a material, ideational or institutional nature. This theme will be further explored in the next part.

11

The Functionalist Approach

Functionalism is to an extent interwoven with structuralism, insofar as it attempts to explain the functions of countless social phenomena and their structures, though not necessarily following Malinowski or Levi-Strauss. Social anthropologist Leach (1970:9) writes:

> I myself was once a pupil of Malinowski, and I am at heart still a 'functionalist' even though I recognize the limitations of Malinowski's own brand of theory. Although I have occasionally used the 'structuralist' methods of Levi-Strauss to illuminate particular features of particular cultural systems, the gap between my general position and that of Levi-Strauss is very wide.

What distinguishes functionalism is its emphasis on 'need'. As Freud remarked to his friend Ferenzci (Jones: 1961:42): 'Our intention is to place Lamarckism entirely on our basis and show that "need", which creates and transforms organs is nothing other than the power of unconscious ideas over the body.' As it happens, Malinowski, one of the founders of functionalism, fully agreed with this point of view. In an article in *Psyche* (1923:293–322) entitled 'Psychoanalysis and Anthropology' he wrote: 'By my analysis, I have established that Freud's theories not only roughly correspond to human psychology, but that they follow closely the modifications in human nature brought about by various constitutions of society.'

The central tenet of the functionalist school, as proclaimed by Malinowski is that:

> Every institution or custom has a function in the society in which it occurs and that all the institutions cohere to make a viable unit, or even organism. This functioning was further the raison d'être of the institution and eventually, by implication the 'cause' of the institution, which by now had come to be regarded as meeting a 'need', biological, psychological or social or a combination of all three. Thus the various customs and behaviors and beliefs of these societies met various needs arising from the essential nature of men living in association. The needs were those of the group here and now, metabolism, reproduction, bodily comforts, etc.

...and the cultural responses to these were, according to Malinowski (1922, 1944):

> commissariat, kinship, shelter, etc., which comprised the customs and institutions and beliefs of the group. These were of here and now and not dependent upon, or the consequences of, past situations, historical circumstances, for their existence or their characteristics.

Lewis (1976:55) comments: 'The functionalist method which Malinowski so strenuously championed amounted in fact to little more than acknowledging that every custom or institution, however strange and bizarre, served some contemporary (i.e. useful) purpose.'

Durkheim, another prominent functionalist, developed a theory of religion which identified it with social cohesion: religious beliefs and rituals are understood in terms of the role they play in promoting and maintaining social solidarity', while Radcliffe-Brown argued that religious ceremonial, promotes unity and harmony and functions to enhance social solidarity and the survival of society. Malinowski again, saw religion and magic as assisting the individual to cope with situations of stress and anxiety (Macmillan, 1983:139).

According to Lewis (1976:48), Durkheim went so far as to assert that crime was necessary and useful to society (i.e. had a useful function). The punishment that followed was shown as a collective revenge of society on the criminal, thus symbolically reaffirming and restoring the moral values and common loyalties which the criminal had desecrated: 'It was a symbolic lynching in which an outburst of punitive indignation healed the injuries which the criminal had inflicted on society.' Durkheim implied that a certain amount of crime might even be necessary to keep society in a healthy state; thus elevating criminals to the status of public benefactors. These and other functionalist views greatly resemble those of Dr Pangloss, in Voltaire's *Candide*, whose philosophy was based on the contention that we live in the best of all possible worlds, with everything connected and arranged for the best. This is not necessarily an exaggerated comment on functionalist ideas, in so far as Durkheim even maintained that 'war and nationalism were good, because they stimulate individuals to display their most heroic virtues, thereby creating unity and purpose' (I. M. Lewis, 1976:53).

It is in this manner that the functional 'need' (or usefulness) of such social elements as kinship and lineage structures, magic and religion, and even crime and war is explained; the guiding motif being allegedly the achievement of social cohesion and harmony. This principle of functionalist 'need' was also introduced into material culture, with some functionalists, maintaining that occupations such as the fashioning of stone tools, spinning and weaving, pottery and agriculture, and even civilisation itself came about because they were needed when societies reached certain stages of development.

Generally, the functionalist principle of 'need' can be refuted by noting firstly that (contrary to the proverb), 'Necessity is not always the mother of invention', as even an obvious need will often remain unfulfilled due to blinkered imagination. Secondly, even if a useful innovation is made it may frequently be ignored, due to the human tendency to adhere to traditional ways.

There is no better example than that of the Fuegian inhabitants of the Tierra del Fuego at the southernmost part of South America. When Darwin encountered them in the last century he found them stark naked, despite living in a climate which was harsh and inhospitable, and despite the obvious need to protect themselves from the seasonably wintry weather with some covering. Furthermore, at one time or the other they were in contact with neighbours who wore protective clothing but the Fuegians did not follow their example. This is certainly a case where the principle of 'need' was most blatantly ignored.

Since the art of making fire dates back at least $1^1/_2$ million years, it ought to be expected that all present-day people should have acquired this skill. However, according to Forde (1968: 141), the Bora and Japura people of Brazil, as well as the Mbuti of the Congo were, before they knew of matches, ignorant of methods of making fire.

In a similar vein, pottery-making is a worldwide craft known perhaps for 10,000 years. Yet the Blackfoot Indians of Central North America had no pottery until a few years ago, and used hide-pouches in place of cooking pots (Forde, 1968:63). Equally, the Mbuti, as described by Turnbull, were devoid of any knowledge of pottery making.

Other examples where functionalist theory founders is the loss of useful skills. Forde (1968:430) has pointed out that rice and millet cultivation was known by some of the ancestors of the Polynesians, but it was later abandoned. And he adds: 'This would be no more surprising than the probable abandonment of both potmaking and weaving.'

W. H. R. Rivers who did extensive fieldwork in the Pacific wrote in great detail about the abandonment of canoe making in the Torres Islands, off N. Australia (1926:191/196); and the loss of pottery making in Polynesia; and that of the abandonment of the bow and arrow as a hunting weapon in many parts of New Guinea, especially among the people who speak proto-Melanesian languages. Bow and arrow are now only used as toys. It was this comprehensive study of *The Disappearance of*

Useful Arts which persuaded Rivers to reject the 'independent invention theory in culture'. Rivers caused a sensation in 1911, when in an address to the British Association on the topic of 'The Ethnological Analysis of Culture' he announced his conversion to the historical, diffusionist interpretation of what he called primitive cultures. 'The choice in ethnology,' he said, 'is between accepting diffusion, or the confusion of Bastian.' Far from his diffusionist conversion diminishing his prestige, he was called upon in 1922 to assume the presidency of the Royal Anthropological Institute in London, still considered the highest honour in British anthropology. Unfortunately, he died suddenly before taking office. Had he survived, British anthropology might most likely have taken a different course. Instead it degenerated (up to the present) into an anti-diffusionist cul-de-sac.

To quarrel with functionalism one need not deny that social phenomena can have beneficial functions, just as most products of material culture are useful and others are not. Objectionable are some of the means by which functionalists try to explain the independent origin of identical cultural phenomena and products, when found in distant parts of the world, attributing them to psychological factors common to all mankind i.e. Bastian's psychic unity'.

Malinowski was a leading exponent of such thinking. In his book, *A Scientific Theory of Culture* (1944), he presents us with a veritable eulogy of Frazer whom he describes as one of the world's greatest teachers and masters on the subject (though he remarks that few of his other theoretical contributions are acceptable today). Malinowski points out that Frazer was essentially addicted to psychological interpretations of human belief and practice (p. 188), and also, 'Frazer believes in the essential similarity of the human mind and human nature. He sees clearly that "human nature" has to be assessed primarily in terms of human needs' (p. 212).

Edmund Leach (in his *Golden Bough or Gilded Twig?*, 1961) commenting on this aspect of Frazer, rightly observed, 'He took over from Bastian the assumption that the fundamental psychology of human beings will be everywhere reflected by similar customary behaviours, or, conversely, that similar customs have always the same symbolic implications regardless of the context in which they appear.'

'A host of writers,' writes Malinowski, 'such as Wundt and Crawley, Westermarck and Lang, Frazer and Freud, have approached fundamental problems such as origins of magic and religion, of morals and totemism of taboo and mana, by propounding exclusively psychological solutions.' Furthermore, according to Malinowski, 'Freud took his anthropological evidence from Frazer.' Neither does Malinowski forget E. B. Tylor (the father of unilinear cultural evolutionism), noting that 'Tylor's minimum definition of religion and his whole theoretical concept of animism was that the essence of primitive faith and philosophy, was primarily psychological.'

And while Malinowski shows little enthusiasm for Frazer's other, purely theoretical contributions (p. 19), his views about the psychological origin of cultural phenomena, matches those of Frazer, Tylor and Freud. In the following we recall (see p.68) Freud's words on the subject and Malinowski's compliance with them. Freud: 'Our intention is to place Lamarckism entirely on our basis and show that "need", which creates and transforms organs, is nothing other than power of unconscious ideas over the body . . .'; and Malinowski: 'By my analysis, I have established that Freud's theories roughly correspond to human psychology, and that they closely follow the modifications in human nature brought about by various constitutions in society.'

What we find here is nothing less than a complete surrender to Bastian's principle of the 'psychic unity of all mankind'. Again the matter is explainable only on the basis of Lamarckian inheritance prevalent in an all-pervading unilinear cultural process. As shown above. Freud endorsed Lamarckism (now considered erroneous) and its part played in social theory. On the other hand, most modern social anthropologists, while continuing to use the same Freudian approach in their social theories tend to overlook its Lamarckian basis, thereby clinging to outdated concepts.

12

The Concerns of Social Anthropology

In the strict sense of the meaning, 'society' or 'social life' as dealt with by contemporary social anthropology is a relatively recent historical phenomenon. It began when culturally developing human groups formed sedentary communities exceeding in numbers the extended families or bands of hunter-gatherers. What we find is that humans, after about four million years of a culturally more or less simple hunter-gatherers' existence, show the first appearance of sizeable social units in a mainly agricultural context about 10,000 years ago.

It is likely that even earlier Neanderthal cave, or tent-dwellers, formed already what may be called 'tribal societies'. Equally so, large scale pre-agricultural sedentary food-collecting settlements may have existed in Jericho and elsewhere. But we know nothing of either's internal social organisation or customs. On the other hand, the societies, clans or tribes which form the main study objects of social anthropologists are hardly older than 6,000 years, if that. Macmillan's *Student Encyclopaedia of Sociology* (1983), confirms this since none of the subjects indexed has any connection with a pre-agricultural mode of life. The terms hunter-gatherer, or gatherer-hunter are not even listed, and the book's sociology or social-science is almost exclusively concerned with sedentary societies.

A close study of culturally simple hunter-gatherers, in their natural state (as yet little influenced by their culturally more advanced neighbours), finds mainly simple family ties, which as in any healthy family today, are based on love, affection, companionship, and above all, sharing. This does not exclude the possibility of infanticide among primitive hunter-gatherers, nor that of internal conflicts (even violent ones), and even aggression. But field workers who have lived among such primitive groups have confirmed that such incidents are rare and disagreements when occurring are usually amicably resolved. However, such pristine conditions mostly ceased when some of these groups or bands adopted agriculture. R. Leakey (1981:227), has pointed out that the principal reason for the Khoi's (S. African food-gatherers) gradual adoption of agriculture is due to pressure from the government. An increased food supply led to the larger social groupings of tribal societies, villages and towns and eventually states and nations. In such novel situations, new sets of personal relations (now called social relations) had to be arbitrarily arrived at to keep order, representing an entirely novel and artificial state of affairs, not being quite compatible with man's basic instincts and aspirations, which biologically seen are more suited to a simple family life within the confines of small groups.

Elsewhere we endorse the view that civilisation arose from special historical events connected with big river irrigative agriculture; and that the corresponding social situation led to the emergence of kingship, godhead and organised religion, followed by a host of subsidiary institutions, customs and rituals.

It is surmised that this ancient Egyptian or Sumerian model of 'urban literate civilisation' with all its ramifications, subsequently became the catalyst from which later civilisations, and their client groups, drew their cultural, and even more, their social heritage, although with adaptations and transformations to fit new conditions. In addition, the system found its imitators in the more culturally backward food-producing tribal societies forming on the fringes of early civilisations. Here the original

customs and institutions derived from higher cultures underwent many changes, and in their new adulterated state led
to aberrations in customs and behaviour, which nowadays seem
to defy logical explanation. Yet many modern social anthropologists ascribe these aberrations to inborn tendencies of a
psychological nature and ignore the very real possibility of
historical explanations. In any case, these lower cultured semi-
civilised tribal groups, surviving in various forms into the
present, have become the favoured hunting ground and object
study of social anthropologists.

To sum up – social organisation, going beyond extended
family groups or bands of hunter-gatherers, is arbitrary, and not
necessarily logical or common sense. It stems from historical
association, thereafter being passed on from generation to
generation, by habit, tradition or by simple inertia, sometimes
with no apparent benefit or purpose. Its expressions include
initiation ceremonies, human and animal sacrifice, cannibalism,
witchcraft, sorcery, totem and tabu, couvade, the seclusion of
young girls for months and even years at the advent of puberty,
ritual murder, the extension of earlobes, the chiselling of holes
into the 'living' skull (trepanning), the ritual removal of an
adult's front teeth, and the barbaric act of female circumcision.
The fact that these and other social customs find their parallels
in many parts of the world, points to common historical
sources, a conclusion which will be more closely examined in a
sequel to this treatise.

It is precisely because such parallels have been ascribed to
inborn human psychology (the psychic unity of all mankind),
that social theories of unilinear and multilinear evolution,
functionalism and structuralism, and other conceptions, inspired
by Bastian, Freud, Frazer, Tylor, and taken up by Levi-Strauss,
Malinowski, Steward, Glyn Daniel, and others, have been
concocted.

It would even seem that human expansion beyond family
groups or bands of food-gatherers and the subsequent attempt
to cope with the complexities of social life in an ever expanding
world population have added unprecedented difficulties to
human biological existence. This has now culminated in a

world of separate sovereign nations plagued by frequent conflicts and wars. The crediting of these new and larger social
assemblages (which are hardly 6,000 years old) and thus represent less than $1/600$th part of total human existence, as being
ordained by fixed social laws, is one of the great paradoxes of
so-called social science. Even more so, the idea that these
artificially assembled social groups, the family-plus up to the
complex of western civilisation, are ruled by social laws which
stem from an inborn psychology, is an almost pathological
obsession currently prevalent amongst large sections of social
anthropologists.

13

The Syndrome of Cultural Under-Development

What do we mean by cultural under-development? It is well known that apes are incipient tool-users and that we can perhaps speak of a rudimentary ape-culture, even involving the teaching of their young. Furthermore, the ape's cultural aptitudes cannot be considered a novel acquirement and probably have persisted unchanged over many millions of years. Some hominids have expanded on this by developing stone-tool technologies. These became increasingly more elaborate, passing through the stages of simple pebble-tools (Hadar and Oldowan), followed by the Acheulian, Mousterian, Upper Palaeolithic and Neolithic sequences, with all their known variations and refinements. It appears, however, that only a limited portion of hominids partook in these lithic developments, *Homo sapiens* included (see Part One, Chapters 6 and 7).

The question is, why should members (or sections) of the biologically uniform species *Homo sapiens* have taken two different cultural trajectories; one portion remaining culturally under-developed for hundreds of thousands of years and the other progressing culturally through various stages, climaxing in modern civilisation.

In broad terms 'cultural under-development', as applied

here, refers to the bulk of those hunter-gatherers' groups, contemporary as well as prehistoric, who still live, or once lived, below the cultural level of agricultural and stock-rearing societies. Writing about the economy of prehistoric hunter-gatherers Geof Bailey (1983: Foreword), points out that human societies have spent over 99 per cent of their cultural history as hunters and gatherers. Others, including Bitchery (1972) and Gowlett (1984), have expressed similar views. Of interest also is Richard Leakey's opinion (1981) that early hominids were perhaps more gatherers than hunters, and therefore he proposed to call them 'gatherer-hunters', while George F. Carter called them 'foragers'.

What is significant is that these culturally less complex hominids roamed the globe for the best part of four million years, up to c. 10,000 years ago, when village life and farming began to make inroads on their hunter-gatherer existence. Remarkable is the fact that some hunter-gatherer populations survived into our age, retaining a lifestyle which, in some respects, resembles that of their prehistoric forebears. During their large timespan cultural progress was minimal.

Until about 50 years ago such survivors were said to have included the Mbuti of the Congo; Kung-San (Bushmen) of South Africa; Veddahs of Ceylon; various jungle tribes of S. India; Semang of Malaya (Australoid); Andaman Islanders; Kunu of Sumatra; Punan of Borneo; various Eskimo tribes of the northern hemisphere; Paiute of Nevada, Utah and Arizona; Indigenes of California; inhabitants of the Tierra del Fuego and assorted Australian aborigines; etc. These examples are of the most diverse racial stock and represent every possible climatic region.

It would, however, be a mistake to lump all those described as hunter-gatherers into one single cultural entity. Although they cannot be strictly separated from one another culturally, it is convenient to divide them into three broad categories.

Into the first category I propose to include those who have been least affected by their sedentary or pastoral neighbours, either by a refusal to become assimilated, or due to less accessible habitats. Among those probably least affected by outside influences are the Mbuti dwarf people, or pygmies of the Congo and the lesser known dark-skinned forest-dwarfs of Malaya, the Kubu of Sumatra, the Punan of Borneo, and the Semang of Malaya, to mention just a few. As to their material culture, their mere survival into the modern age compels us to conclude that their way of life was highly adapted to their environment and in no way inferior in survival value to that of their culturally more complex contemporaries.

R. Lee, (cited by R. Leakey, 1981), known for his extensive studies of hunter-gatherer societies, offered that as a general rule foraging people on the whole deploy tremendous skill and only minimal technology in exploiting their environment. 'They live in such a harmonious relationship with nature that the hunter-gatherer way of life can be described as the most successful adaptation man has ever achieved.' Edmund Leach, goes even further. He wrote (in correspondence) that 'nobody in their senses would live in any other way if they had the space'.

The second category includes hunter-gatherers already more deeply affected by foreign influences. Prominent among these are the Kung-San (Bushmen) of Southern Africa. However, they still reveal ostensibly authentic facets of a sociologically and ideologically unpolluted lifestyle reminiscent of the first category.

Category three comprises that multitude of hunter-gatherers who have already accepted a great part of the way of life of their sedentary agricultural neighbours. They include the Veddahs of Ceylon; Andamanese aborigines, the Sakai and Jakun; the Australoid neighbours of the Semang, a proto-Malay people; certain Negrito people of the Philippines. In the Americas they include peoples such as the Dene, Salish, Paiute, and the Red Indian peoples of California; further south there are numerous Brazilian peoples, while in the Chilean parts there still linger the Fuegians and their close neighbours, the Chonoans and the Onas. In the arctic regions we encounter various Eskimo peoples of different complexion with an economy partly based on fishing and hunting. All of these above mentioned can only nominally be considered as genuine (bona-fide) hunter-gatherer people. Desmond Clark comments (*Man the Hunter*, 1968):

'Most of the hunter-gatherer groups existing today are living in contact with more complex societies and technologies. They can therefore no longer be considered "typical" or useful for any comparison with prehistoric populations.' I suggest that some of those listed in categories one and two above, are exceptions to this generalisation.

By occasionally describing the first category of hunter-gatherers as cultural primitives or even ultra-primitives, I try to contrast them with the more acculturated ones belonging to categories two and three. It needs stressing that the foregoing description does not imply an inferior type of culture or intellect but one that is less complex or sophisticated. At the Third International Conference on Hunter-Gatherers (Carmen Shrire, 1983) a majority view concluded that there were two different types of hunter-gatherers, primary and secondary ones. The primary type embodies those groups who do appear to be the aboriginal, autochthonous peoples of their respective regions of whom there exist at present no cultural, historical, or linguistic evidence of a prior agricultural adaptation; while the secondary type appears to derive from sedentary agricultural peoples representing essentially a re-adaptation.

14

Hunter-Gatherer Life

Some general facts about hunter-gatherers, particularly their division into three different cultural categories owing to a lesser or greater complexity or sophistication in their lifestyle, has been dealt with in preceding chapter. There the main concern was with the material and technological aspect of hunter-gatherer life; here the focus will be more on the social and ideational content of their life.

R. Leakey (1981) has pointed out that nineteenth-century anthropologists viewed hunter-gatherers as fossilised societies; primitive savages who had somehow slipped unnoticed and unnoticing into the modern world. Leakey thinks that this view is nonsensical. The hunter-gatherers were as modern in biological terms as the explorers who discovered them. They just happened to sustain themselves by an ancient method. Anthropologist Marshall Sahlins (as quoted by Leakey) has argued that Western anthropologists must not impose Western, that is, materialistic, ethics on their subjects. Sahlins refers to the different goals of the various societies: the pursuit of wealth, property and prestige in the one, and something totally different in the other. He goes so far as to suggest that the hunting and gathering way of life is the original affluent society in which all the people's wants are easily satisfied. As it happens, the hunting and gathering economy is not an incessant search for food, as many anthropologists have supposed, but a system that allows a great deal more leisure than is possible in either agricultural of industrial society.

Some details about culturally primitive food-gathering groups deserve recall in this context. Schebesta (1929:53) has described the 'Negritos' of Malaya as: 'primitive relics, almost, of a prehistoric age, and, "savages" by name only. Their affection towards each other was strong and their little camp had an air of contentment. Their fundamental unit is a small group of relatives. There is no chiefly class, no tribal unity.' Another group, the Kubu of Sumatra, were observed by Forbes (1885: 232) to be as peaceful and having no social classes. He says that until a few years ago they wore no clothes, and concludes that: 'one cannot help feeling that they are harmless, overgrown children of the woods'. Others like the Punan of Borneo were – culturally speaking – among the most primitive people in the world. Hose and McDougall (1912:180/185), described them as having no social classes, no houses, and their property was held communally. Harmony and mutual help were the rule within their family circle, as well as throughout their larger groupings; the men generally treated their wives and children with kindness and the women performed their duties cheerfully and faithfully. Schebesta spoke of the Semang of Malaya (1929: 279) as having no tribal organisation; while Forde observed that they were unable to fashion even simple pebble-tools (1968: 70).

Many social anthropologists hold the view that the allegedly peaceful and rather idyllic lifestyle of culturally primitive food-gatherers (past and present) is more myth than reality and that they have been just as aggressive and superstitions as their materially more advanced agricultural and pastoral neighbours. In this respect the remarks of Mayer Fortes at a Symposium (1973:431) are typical: 'No human group has yet been dis-covered which is devoid of some form of extra social organisa-tion, leadership, ritual practices, or aggressive tendencies.' In support he referred to the field researches of Radcliffe-Brown, among the Andamanese; the Seligmans among the Veddahs of Ceylon; and the field researches of Father Wilhelm Schmidt of the Vienna School from 1906 onwards. A study of the Mbuti and similar food-gathering groups, which I shall cite in due course, will serve as a rebuttal of these views.

However, in certain respects, Fortes's observations are valid for those surviving food-gatherers and hunters listed in category two (see Chapter 13) who have been greatly affected by their contacts with semi-civilised and civilised populations. The recent history of the Khoi-San (so-called Bushmen) of South Africa, commented upon by Dorman (1925:43) is a case in point. 'Although the Bushmen had already been driven from their original homeland in the Cape by advancing Bantu-speaking Negroes, they could still be described on the whole as happy, lively people, full of merriment ... though when opposed or thwarted they could become savage, cruel and vindictive ... When the Dutch arrived at the Cape at 1650, they slaughtered the game on which the Bushmen lived and took away their land. In self-defence the latter took to looting the Dutchmen's herds, and in retaliation were shot like wild beasts, while the women and children were carried into slavery (1925:205/206).' Despite these ordeals, the Kung (the section of Bushmen who now occupy part of the Kalahari desert) have preserved some of their original ethics.'

Thus, Lee (cited by R. Leakey, 1981): 'Sharing of resources deeply pervades the behaviour and values of Kung foragers within the family and between families, and it is extended to the boundaries of the social universe. This ethic is not confined to the Kung; it is a feature of hunter-gatherers in general. In the same vein as the sharing ethic, comes a surprising degree of egalitarianism. The Kung have no chiefs and no leaders ... Disputes are readily defused through communal bantering. No one gives orders or takes them.'

Further afield in Asia the fate of some food-gathering tribes of India was reported by Buchanan (1867). He travelled through Mysore, Canara, and Malabar, and observed that: 'In nearly all cases they have been subjected to slavery at the hands of their neighbours, and in their service have learned the rudiments of agriculture and other arts.'

These and countless other instances show that *c.* 6,000 years of contact with semi-civilised and civilised populations has inevitably tainted and corrupted the cultures of these formerly ultra-primitive food-gatherers and saddled them with many of the customs and social attributes of sedentary tribal societies. These latter acquirements can be enumerated as: hierarchical status relations with chiefs, headmen and councillors; complicated kinship rules, mythical accounts; totems and taboos; rituals of initiation, marriage and burial; ancestor worship; superstitious beliefs in magic, witchcraft and sorcery; trepanning, circumcision, cannibalism, human sacrifice; all kinds of religious practices and more. It is surprising that many of these culturally primitive groups have not become even more influenced in their basic nature and life style by such extensive cohabitation. Thus we find some surviving examples (among them most prominent the Mbuti pygmies of the Congo) who until some decades ago had remained relatively free from such social incursions.

These latter groups had no chiefs or headmen, no councils or legal systems, no beliefs in supernatural sanctions, and practices of magic, witchcraft or sorcery. There were no initiation rites or complicated kinship systems and when left undisturbed their life was peaceful and non-aggressive.

The significance of the Mbuti of the Congo as they were found by Colin Turnbull in the 1950s rests not so much on the material aspect of their culture, but on its ideational content. In revealing their basic mentality with its no-nonsense attitude to foreign belief-systems and social attitudes, Turnbull has made a major contribution to the ethnological sciences, which given due recognition, could greatly affect current ideas prevalent in social anthropology. He discovered that the Mbuti still had (or almost have) no tribalistic ideology or philosophy, although they had been in reported contact with the outside world for thousands of years. Earliest mentioning of the Mbuti of the Congo comes from a letter of a Pharaoh of the Sixth Dynasty (c. 2,360 BC) to his general Hechuf who brought a dancer back from a Congo expedition. The Mbuti pygmies are also mentioned by Herodotus (c. 500 BC), and Aristotle (c. 400 BC) (African Pygmies, 1986:346).

Thus, writes Turnbull, in the *Scientific American*, (January, 1963): 'After centuries of contact with the "more advanced" cultures of the Bantu villages and in spite of all appearances, their acculturation to any other mode of life remains almost nil.' And again in *African Pygmies* (1986): 'The pyramidal edifice of family, lineage and clan does not exist in any form. There is no intermediate grouping between family and band, and none between band and total population ... The solidarity of the band is the prime factor in ensuring group survival, in which kinship once again plays a minimal role.' Also according to Turnbull (1966), and I quote at length:

> The overall picture of the present Mbuti culture is one of a society where the lack of formal structure is so evident that one wonders why there is no complete disintegration. There are not only no chiefs or councils of elders, but no ritual specialists, and no lineage systems, and no body of beliefs in super-natural sanctions. The only effective political unit is the band, which can be defined as a unit of Mbuti sharing and recognizing a common hunting territory. Membership of the band is fluid.

Kinship:

> Neither men nor women are able to name their grand-parents (nor even name any brother or sister of their grandparents), unless they have a vivid personal recollection of them. The only occasion of any significance to such detail would be marriage.

Initiation and marriage:

> Though manhood is every bit as important to the Mbuti as to the Bantu villagers, there is no formal Mbuti initiation ritual. And although marriage is considered just as sacred, it takes place in the forest in an apparently utterly casual manner, without any ceremony or ritual joining the boy and the girl or their family ...

Death is accepted as a natural event and does not lead to any accusation of witch-craft or sorcery, except in a Bantu village context. Burial is performed simply and quickly and excessive display of grief is frowned upon. The forest procedure (of natural death) would simply be to bury the body, then abandon the camp and build a new one, never again mentioning the dead person. The Mbuti boys have no fear of breaking all the initiation rules imposed by their Bantu masters when they are absent. Under certain conditions they will adopt the Bantu's burial customs, because 'they (the Bantus) will give them much food'.

As to the Mbuti's Belief-System Turnbull observed, on magic:

The Mbuti have no magic with which to counter magic, and no witchcraft to counter witchcraft or sorcery (1977: 54). As to witchcraft seen as a malevolent sorcery, they cannot believe that a 'bundle of sticks' is going to afford much help (1966:59).

On religious afterlife, the Mbuti say:

that to try and go into the future is to walk blindly, and their response to their Bantu masters and missionaries, or any who claim knowledge of afterlife, is: How do you know, have you died and been there before?

The supernatural:

Among the Mbuti there is a belief in a power greater than themselves, which is not of the natural order or of the world as they see it around them. They explain these unknown forces with a rather confused terminology; the Epelu Mbuti use five different terms and these are drawn from Lese, Bira, and Ndaka, which are all settled villages of their Bantu masters among whom they live. One ought to comment here that the vagueness of such a doctrine clearly suggests long indoctrination.

This can also be seen in relation to godhead: The Mbuti strenuously refuse to admit that there is any sense in trying to describe what we call here as "godhead". They will say: 'Unless you have seen it there is no sense in accepting it'.

In judging the discoveries of Turnbull, one must note the rare coincidence of his, finding the Mbuti when they were still in a socially and ideationally fairly uncontaminated state – just at the eleventh hour – before their almost inevitable demise as a cultural prototype. It is difficult to recall any similar discovery taking place so close to the end of an era.

Decades later, in 1966, 'between 100,000 and 200,000 Mbuti survive in the African states the Cameroon, Congo, Gabon, the Central African Republic, Zaire, Burundi, Equatorial Guinea, and Rwanda. As a people of the forest, the Mbuti have been rejected, ignored and exploited. Some have left their native milieu, tempted by the prospect of an easier life. They have inevitably encountered money, alcoholism and diseases they never knew before. It is a kind of life which alienates and destroys them.' (From a film review of *Pygmies*, in the *Bangkok Post* 11 September 1986.)

Most social anthropologists consider the kinship syndrome an axiomatic and universally applicable feature in the evolution of societies. Its virtual non-existence among the Mbuti and some other culturally 'low-level' hunter-gatherers negates this assumption. While the Mbuti show an almost total absence of supernatural ideas, it is a fact that they acquired some confused notions about a superior power. However, considering their rejection of the notions of a 'godhead' and of a 'religious thereafter', it is likely that such precepts were assimilated during the Mbuti's long exposure to culturally more complex societies. Such a partial acceptance of alien influences can hardly be attributed to the Mbuti's having receded from a former socially and ideationally more complex stage. On the contrary, given the common-sense attitude generally exhibited by the Mbuti, we can assume that before they became saddled with such ideatic burdens they could only have been ideationally less adept than their modern descendants.

In Asia – thousands of miles apart from the Mbuti of the

Congo – we have examples of hunter-gatherers resembling
them in some aspects of their mentality, though to a lesser
degree in their total rejection of paranormal ideas and the
absence of social conventions. Although living in similar conditions to the Mbuti these people have been more subject to
foreign influences. Among other less typical examples are the
African Khoi-San (Bushmen) of the Kalahari desert in South
Africa.

From the example of the Mbuti we can presume that the
mentality of prehistoric hunter-gatherers as a whole, in respect
of social customs and their world of ideas, could only have been
less complex ideationally than that of their presently surviving
culturally ultra-primitive successors (i.e. those listed in our
category one). Although the available evidence may serve as a
general indicator of the ideational status of hunter-gatherers
the world over during the preceding hundreds of thousands of
years, before food-producing societies became prominent, there
is no direct evidence to support this claim. On the other hand,
the circumstantial evidence is overwhelming.

In material culture, similar analogies are possible, between
modern and prehistoric hunter-gatherers. Yet as in the above,
they are largely based on circumstantial evidence. On the other
hand, many anthropologists assert that this type of analogy is
not feasible. Commenting on a previous draft of this paper,
social anthropologist Edmund Leach writes (personal communication): 'A substantial portion of the authors you mention
believed that ethnographic evidence could be used to reconstruct the history of the distant past. You yourself seem to believe
this even though you are sceptical concerning most efforts of
this kind. I myself would agree with those other anthropologists
(with whom most serious historians would also agree) who say
that it is impossible to reconstruct history on the basis of
circumstantial evidence.'

Similar views have been expressed by Gowlett (1984). He
writes: 'Any direct analogy between modern hunter-gatherers
and societies represented in the archaeological evidence would
be unwise ... Analogy is especially suspect when the practice of one modern people is invoked to explain apparently

similar behavior in the past, for usually there are many other possibilities.'

In spite of these objections I maintain that a large degree of comparison between the material cultures of prehistoric and modern hunter-gatherers is possible. In this respect, the cultural propensities of *Homo habilis*, as summarised by Tobias (1983), serve as a useful starting point.

15

The Culture Complex of Homo Habilis

Based on P. V. Tobias's studies

Doubtless, the cultural achievements of even the earliest hominids may go far beyond the abilities we can assess from their crude stone-tool remains alone. Yet archaeological remains other than stone, pointing to such hidden capacities, are rare due to the perishable nature of the material they may have used.

However, partly by inference I propose that P. V. Tobias's (1983a) summary of the cultural aptitudes of *Homo habilis* cited below is one of the most impressive statements so far made about the cultural beginning of this first *Homo* species, between *c.* 2.3 and *c.* 1.7 million years ago.

P. V. Tobias writes that the cultural assemblage of *Homo habilis* have been designated as Oldowan (M. D. Leakey 1971). Among the stone tools there was a predominance of choppers of which five types, side, end, two-edged, pointed and chisel-edged have been described. Other forms included proto-bifaces, polyhedrons, discoids, spheroids, heavy-duty and light-duty scrapers, burins and sundry other tools. 'To this variegated suite of tool-types must be added the evidence that *Homo habilis* was capable of constructing some form of shelter in the form at least of stone-walling. The implemental and constructional activities bespeak a complex culture.' Evidence suggests that 'the culture of *Homo habilis* included the aimed throwing of missiles, the butchery of large animal carcasses with stone tools, the transport of meat and other foods to a home base, delayed consumption, the sharing of food and the distribution of the meat to adult and juvenile members of the group (M. D. Leakey, 1971; Isaac, 1978). All in all, the cultural achievements, both those observed and those inferred imply a great deal of intelligent activity and could hardly have taken place without the assistance of some form speech.'

It is noteworthy that we find most of the activities outlined above (with the exception of *Homo habilis*'s elaborate stone-tool technology), incorporated in the lifestyle of many modern hunter-gatherers, with the addition, of course, of intrusive cultural elements derived over the millennia from culturally more complex neighbour populations. Consequently, considering modern hunter-gatherer populations of a primitive cultural level (my categories one and two), I conclude that during the more than two million years which have elapsed since *Homo habilis*'s beginnings, their cultural evolution (read development), was practically nil. And yet modern hunter-gatherers of a low cultural level are fully fledged *Homo sapiens*, sharing with us an average brain capacity of *c.* 1,350 cc., which is more than double that computed for *Homo habilis* (of 646 cc.), a figure quoted by P. V. Tobias (1983a).

The comparative picture becomes particularly pertinent if we use the standard of stone-tool technology as a cultural and intellectual parameter. In this respect, the Mbuti (without any stone-tool making propensities) would have to be considered as ranging culturally and intellectually below *Homo habilis* of more than two million years ago. Yet, as the observations of Colin Turnbull show, the Mbuti cannot be considered intellectually inferior to contemporary civilised populations, though they lag behind in cultural complexity and sophistication. Observes R. L. Holloway (1984): 'Were modern living human hunters and gatherers to be judged on the basis of stone-tool technology alone, they would probably be considered less advanced "brain-wise", than Neanderthalers.'

Among the Mbuti, Turnbull informs me, their economy requires a minimum technology, still at a stone-age level. In the use of indigenous tools (other than stone), they are highly adaptive and adaptable, but the degree of their adaptation is light by the very nature of their life and environment which has remained stable for many thousands of years – and may, under similar environmental conditions have resembled in many, though not in all respects, that of prehistoric hunter-gatherer populations living during most of the millions of years of the stone age. The most conspicuous prehistoric achievements of these latter groups being the art of the making, and the use, of fire and the use of bow and arrow; besides possessing, most importantly, the full capacity for articulate speech as evolved in *Homo habilis* and perfected in *Homo sapiens*.

In adjudging the Mbuti's intellectual level, Turnbull remarks, 'My observations in this field were that in intelligence many Mbuti seemed to have a far higher I.Q. than I, and some were exceptionally dumb in some respects and exceptionally bright in others. The kind of variation I may expect in any population I happen to know. All of them seemed to have far greater knowledge and understanding of the world they lived in than most of us in our complex western world.'

16

Conclusions

The purpose of Part Two has been to show that the evolutionary (psychological) interpretation of cultural phenomena is outdated. In contrast, an evaluation of accumulated fossil evidence particularly in relation to its cranial aspect (for more details see Part Three) points to a historical diffusionist interpretation.

In examining current anthropological theories such as cultural multilinealism (which can be shown to be merely another form of discarded unilinealism), structuralism and functionalism, we find them all linked to the now outdated concept of cultural evolution, which in turn is again based on the psychologically linked thesis of the equal working of the human mind. This latter concept was originally formulated by Bastian and further developed by Morgan, Tylor, Frazer, Freud and others. Furthermore, all of these theories are rooted in Lamarckian heredity, an observation which finds prominent exposition in the ethnological writings of Sigmund Freud, who admits being a convinced Lamarckist.

What needs stressing here is that conventional social anthropology which draws its study material almost exclusively from sedentary societies (while disregarding hunter-gatherer history), is principally linked to the above mentioned cultural evolutionary theories. Yet simple hunter-gatherer societies have occupied our earth for four million years or more, some of them even surviving into our age while in contrast, the more culturally complex agriculturally based sedentary societies are at best only 10,000 years old, representing thus a mere one quarter per cent of total human history, or two per cent of the duration of *Homo sapiens*'s earthly presence, whose age is about 300,000 years.

The rare example of the surviving, culturally ultra-primitive Mbuti hunter-gatherers of the Ituri Forest of Zaire, gives us a possible clue to the lifestyle of hunter-gatherers throughout the preceding four million years of human pre-agricultural existence. Their social life has been described by Colin Turnbull as being free of most of those social customs and institutions which have become part and parcel of more complex sedentary societies, principally those of a tribal character. These customs and institutions can be enumerated as: hierarchical status relations with chiefs, headmen and councilors; complicated kinship rules; mythical accounts; totems and taboos; rituals of initiation, marriage and burial; ancestor worship; superstitious beliefs in magic, witchcraft and sorcery; trepanning; circumcision; cannibalism; human sacrifice; all kinds of religious practices and more. As to their communal behaviour, the Mbuti, when remaining undisturbed by external influences, lead a perfectly peaceful existence, and show no aggressive tendencies.

In comparison, the more complex sedentary tribal societies, during their relatively short history, have become saddled with the above mentioned medley of social customs and institutions, and are prone to aggression as well as being riddled by internal and external conflicts. Also worth mentioning is that the conditions which apply to human total history, also logically apply to the entire history of *Homo sapiens*, whose total food-gathering sequence stretches over 300,000 years (minus the last 10,000 years of sedentary life, apart from some surviving hunter-gatherers). Fossil evidence shows that during *Homo sapiens*'s entire lifespan, which includes modern humans, there occurred no significant biological change either in their physical or mental makeup. This means that also their psychological aptitudes must have remained basically unchanged during the entire period.

It is therefore inconceivable to accept the notion that while the psychological aptitudes of *Homo sapiens* remained unchanged over 300,000 years (minus the last 10,000 years), there could have occurred during the last 10,000 years a tremendous psychological upsurge, leading to the spontaneous worldwide emergence of the above mentioned medley of customs and institutions, all of them being allegedly induced by a cultural evolutionary process, this being a totally deterministic assumption.

The survival of the culturally unsophisticated Mbuti, alone, who are recognised as fully fledged *Homo sapiens* (being free of the above mentioned medley of customs and institutions) already proves the absurdity of the above culturally evolutionary deterministic assertion and its alleged psychological consequences. The further claim that identical cultural parallels found elsewhere in the world, are, and can have been, the result of spontaneously produced independent developments of the same order, as part of an evolutionary process, induced by an inborn psychology, must also be discarded.

I suggest therefore, based on overwhelming evidence, that the rational alternative to outdated cultural evolutionism, is a historically based diffusionist interpretation of cultural events. This further suggests that the emergence of cultural elements and phenomena in human history is in the first place due to local origins (or, in wider aspect, due to specific historical circumstances), while their parallel appearance elsewhere in the world is due (with few exceptions), to a process of spread or diffusion from an original source. And such a process can be postulated irrespective of whether physical contact can be proved or not. A typical example of this is the origin and spread of the great religions of this world, and even more so that of the origin and world wide spread of Western civilisation. The next part to follow will produce additional evidence for the diffusionist interpretation of humankind's cultural history.

Bibliography for Part Two

Bailey, Geoff, 1983, *Hunter-Gatherer Economy in Prehistory*, Cambridge University Press

Buchanan, F. H., 1867, *Journey Through the Countries of Mysore, Canara and Malabar*

Butzer: *see* Caldwell

Caldwell, F. R., 1966, *New Paths to Yesterday*, Thames & Hudson

Cavelli Sforza, L., 1986, *African Pygmies*, Academy Press

Chambers, 1960, *20th Century English Dictionary*, W & R Chambers

Clark, D., *see* Washburn

Daniel, Glyn, 1971, *The First Civilizations*, Pelican (first published 1968

— 1964, *The Idea of Prehistory*, Penguin Books

Darwin, E., 1888, *The Life and Letters of Charles Darwin*, vol. 2, Murray

Darwin, Charles, 1958, *The Origin of Species*, World Classics, Oxford University Press

Dornan, S. S., 1925, *Pygmies and Bushmen of the Kalahari*

Dyer, W. T. T., 1888, 'The Duke of Argyll and the New Darwinians', in *Nature* 247/248

Forbes, H. O., 1885, *A Naturalist Wandering in the Eastern Archipelago*

Forde, D. C., 1968, *Habitat Economy and Society*, Methuen

Fortes, Mayer, 1973, *Elliot Smith Symposium*, The Zoological Society of London

Freeman, D., 1974, 'The Evolutionary Theories of Charles Darwin and Herbert Spencer', (*Current Anthropology* vol. 15 No. 3, Sept.)

Freud, S., *see* Jones, E.

Goodrich, E. S., 1912, *The Evolution of Living Organisms*, Jack

Gowlett, J., 1984, *Ascent to Civilization*, Collins

Harris, M., 1972, *The Rise of Anthropological Theory*, Kegan Paul

Hocart, A. M., 1941, *Kingship*, Watts

Holloway, R. L., 1984. 'The Poor Brain of H. Neanderthalensis', in *Ancestors, the Hard Evidence*, Allan R. Liss, New York

Hose, C., and MacDougall, W., 1912, *The Pagan Tribes of Borneo*

Huxley, J., 1957, *Evolution in Action*, Montana Books, New American Library, New York

Isaac, G. L., 1978, 'The Archaeological Evidence for the Activities of Early African Hominids', in *Early African Hominids*, ed. C. J. Jolly, Duckworth

Joel, C. E., 1970, in *The New Diffusionist* Nos. 1 and 3 Grt Gransden

— 1973, in *The New Diffusionist* Nos. 9 and 3 Grt Gransden

— 1981, in *Historical Diffusionism* No. 33 Dec

Johanson, D. C., and Edey M. A., 1982, *Lucy*, Granada Publishing Ltd

Jones, E., 1961, *The Life of Sigmund Freud*, Pelican

Koestler, A., 1978, *Janus*, Hutchinson

Kroeber, A. L., *see* Daniel, G., 1971

Laufer, B., 1918, *The American Anthropologist* 20, review of R. H. Lowie's *Culture and Ethnology*

Leach, E., 1970, *Levi-Strauss*, Fontana/Collins

Leakey, M. D., 1971, *Olduvai Gorge* vol. 3, 'Excavations in Beds I and II', Cambridge University Press

Leakey, R. E., and R. Lewin, 1979/1981. *People of the Lake*, Collins

— 1981, *The Making of Mankind*, Michael Joseph

Levi-Strauss, C., 1968, *Structural Anthropology*, Allen Lane

— 1962, *The Savage Mind*, Weidenfeld

Lewis, I. M., 1976, *Social Anthropology in Perspective*, Pelican

Lowie, R. H., 1940, *An Introduction to Cultural Anthropology*, Farrar and Reinhart New York

Lukes, St, 1975, *Emile Durkheim, His Life and Work*, Allen Lane

Macmillan's *Student Encyclopaedia of Archaeology*, 1983

Malinowski, B., 1923, 'On Freud: Psychoanalysis & Anthropology', in *Psyche* 4

— 1944, *A Scientific Theory of Culture*, University of North Carolina

Morgan Lewis, 1877, *Ancient Society*, 1964 edition. Edited by L. S. White, Harvard University Press

Morris, B., in *MAN*, Dec. 1974:542/555, R. A. I. Publication

Parker, S. T., and Gibson, K. R., 1979, *A Developmental Model for the Evolution of Language and Intelligence in Early Hominids*, The Behavioral and Brain Sciences 2, 367–381 (1979)

Piddington, R., 1957, 'Malinowski's Theory of Needs', in *Man and Culture*, R. W. Firth (ed), Routledge

Radcliffe-Brown, A. R., *The Andaman Islanders*, Cambridge

Rivers, W. H. R., 1926, *Psychology and Ethnology*, Kegan Paul

Schebesta, P., 1929, *Among the Forest Dwarfs of Malaya*

Shrire, Carmen, 1983, *Past and Present in Hunter-Gatherer Studies*, Academic Press, Third International Conference of Hunter-Gatherers, Bad Homburg, W. Germany, 13–16 June 1983

Smith, G. E., 1973, *Centenary Symposium Report*, Zoological Society, London

Spencer, H., 1857, *The Ultimate Laws of Physiology*

— 1866, *Principles of Physiology*

Seligman, C. G., 1910, *The Melanesians of British New Guinea*, Cambridge

Seligman, B., 1911, *The Veddahs*, Cambridge

Steward, J. H., 1953, *Multilinear Evolution; Evolution and Process in Anthropology Today*, University of Chicago Press

Steward, H. J., 1955, *The Theory of Culture Change*, Urbana, USA *Sunday Times*, 8 October 1972

Tobias, P. V., 1956, *On the Survival of Bushman in Africa*

— 1983, *Recent Advances in the Evolution of the Hominids: With Special Reference to Brain and Speech*, Pontifical Academy of Sciences, Scripta Varia, vol. 50, pp. 85–140

Turnbull, C. M., 1966, *Wayward Servants, The Two Worlds of African Pygmies*, Eyre & Spottiswood
— 1965, *The Mbuti of the Ituri Forest*, Eyre & Spottiswood
Tylor, E. B., 1881, *Anthropology*
Voltaire, 1954 (written in 1756), *Candide*, Penguin Classics
White, L., 1959, *The Evolution of Culture*, McGraw Hill, New York
Wolheim, R., 1971, *Freud*, Fontana/Collins
Washburn, 1968, *In Symposium, Man the Hunter*

Part Three

The Significance of
Cultural Parallels

17

Indications of Cultural Spread

Following Columbus's discovery of America, numerous explorers, naturalists, missionaries and adventurers swarmed over the globe and through succeeding centuries carried home tales of strange customs and institutions of mostly culturally less advanced tribal societies they had visited. Upon examination of their records it was found that even some of the most bizarre of these customs they had encountered had parallels among peoples living oceans apart.

It seems likely that the sifting and analysing of these recorded experiences led to the birth of cultural, and then social, anthropology. These studies blossomed during the period when Lamarckian and Darwinian theories inspired biologically-linked cultural evolutionary theories. On their basis some anthropologists postulated the biological evolutionary process could be applied to cultural phenomena also. Resulting concepts formulated by Bastian, Spencer, Tylor, Morgan, Frazer and others, led to the unilineal theory of independent cultural evolution (later abandoned and superseded by ideas of cultural multilinealism). Unilinealists had assumed that worldwide cultural parallels drawn from less advanced societies were attributable to independent origins. Sir James Frazer in *The Golden Bough* attempted to unify such evidence by assuming that all people had once passed through the same cultural evolutionary phases. This aspect of anthropology has been aptly summarised by Levi-Strauss (1968) – quoted previously – I repeat:

> The evolutionist interpretation in anthropology clearly derives from evolutionism in biology. Western civilisation thus appears to be the most advanced expression of the evolution of societies, while primitive groups are 'survivals' of earlier stages, whose logical classification reflects their order of appearance in time.

The theoretical circumstances surrounding the emergence of cultural evolutionary thinking, have been dealt with at length in the preceding part.

Attention must now be drawn to a fallacy which adherents of the cultural evolution school have consistently ignored.

Cultural parallels found in less advanced tribal societies are routinely (though mistakenly) attributed to independent origins as part of a cultural evolutionary continuum. Yet, cultural parallels between civilised societies cannot be squeezed into the framework of a cultural evolutionary process. The most conspicuous witnesses to this is our own Western civilisation and the world's great religions.

Western civilisation dates roughly from a period of the European conquest of remote parts of our globe accompanied by its cultural impositions which followed Columbus's discovery of America. Most of the creations of Western civilisation can now be found worldwide. Within this enormous cultural upsurge, resulting from the growth of Western civilisation and spread, there are no examples of cultural innovations which can be attributed to a genetically linked evolutionary process. This holds true whether we take the examples of the internal combustion engine, the moon vehicle, television and computers, or dress trivialities such as collar and tie, and the cufflink. It also applies to artistic scientific, philosophical and political ideas.

As to religion, all the great ones originated at specific places and times. Christian beginnings for example, revolve around the person of Jesus Christ, born in Palestine. What is self-evident here is that before Christ's birth there was no such thing as a Christian faith although during its growth and spread many practices from other religions, particularly from Judaism, became incorporated. Today, Christianity is practiced by about 1,000 million people worldwide and wherever it is found it can be proved that it reached there by human contact and that none, even its most bizarre expressions, is attributable to independent origins or related to anything resembling a cultural evolutionary process.

It seems superfluous to stress such a truism, yet when the question of other more primitive religions or belief systems is raised, such as universal animism, sunworship, or divine kingship etc., cultural evolutionists will reject the idea that they can belong to a similar category of religious phenomena, with distinct historical origins, as Christianity, Islam, or Buddhism, ascribing them instead to independent local origins.

The point being aimed at here is that there are thousands of striking examples of worldwide cultural parallels which can hardly be attributed to independent origins. Prominent among them is the origin of American civilisation. Or to take a more widespread example, the worldwide emergence of agriculture. Yet these latter examples (and there are many others), continue to be used in evidence of the independent invention and development theory of culture, serving as a prop for the moribund concept of a Lamarck-based cultural evolution. The main reason being that historical traces for physical contact are either scarce or are altogether missing.

What follows are samples of cultural parallels, which space permitting could be multiplied a hundredfold. Our first case in this scenario is the controversial subject of the origin of American civilisation.

18

America and the Old World

There exist hundreds of pre-Columbian traits in America, both in material culture and in the field of ideas which find striking parallels in the Old World. They indicate intrusions, both from the Atlantic and the Pacific side.

In attempts to explain, contemporary Americanist scholars can be divided into three broad groups. Firstly, those who are fully convinced of the Old World origins of America's Pre-Columbian civilisations, prominent among them R. A. Jairazbhoy, and George F. Carter. Secondly, there is Joseph Needham and others, who, although they demonstrate the presence of numerous, mainly Trans-Pacific, cultural parallels in America, fail to contend that these influences played any decisive part in the formation of early pre-Columbian civilisations. John L. Sorenson a leading Americanist, whose extensive researches in this field demand great respect, occupies perhaps a middle ground between groups one and two. The third group, until recently in the majority, are isolationist scholars, most prominent among them, the late Glyn Daniel (known for his violent anti-diffusionist tirades) and John H. Rowe, who even in the face of the most convincing trans-oceanic parallels denies the possibility of pre-Columbian intrusions from the Old to the New World, excepting those few cases which he attributes to occasional drift voyages by ancient mariners. Ironically though,

Rowe's own large list of trans-oceanic parallels proves just the opposite.

In a symposium, summarised in *Man Across the Sea* (1971), much space has been devoted to the discussion of pre-Columbian and even pre-historic people's capabilities to cross large expanses of water including trans-oceanic and trans-Pacific voyages. The general consensus reached was that based on papers submitted by A. Kehoe, R. A. Kennedy and others; Atlantic as well as Pacific crossings were considered possible from very early times. Kehoe had pointed out that the 'couragh', a very ancient-type boat of sewn-together animal skins stretched over a light wooden frame was probably available to Atlantic Europe from *c.* the eighth millennium BC onwards, while plank boats were present from *c.* 1,500 BC to the late Bronze Age (1971: 277/279). R. A. Kennedy, presenting a paper 'A Trans-Atlantic Stimulus Hypothesis from Mesoamerica and the Caribbean *c.* 3,500 – 2,000 BC, maintains that trans-Atlantic navigational problems can be dismissed as non-existent. He describes the pottery found on both sides of the Atlantic between 3,500 and 2,000 BC as being one of the most diagnostic comparisons, with striking correspondences between the west side of the Atlantic (North and Northwest Africa) and the east side of the Atlantic (South East, Middle and Central America – Caribbean) (1971:267/268).

It used to be generally believed that the original peopling of the Americas occurred *c.* 11,500 BC, when the first prehistoric intruders from Siberia are presumed to have crossed the Bering Strait into Alaska and later penetrated into North America through an ice-free corridor stretching east along the Rocky Mountains. It was assumed that they eventually spread over the entire continent, reaching Patagonia in the extreme south of the continent a few thousand years later.

This theory of an exclusive Bering Strait entry of humans into America is now seriously challenged by various authorities and on various counts. In a summary of recent challenging evidence published in *Man* (March 1988), Ruth Gruhn (University of Alberta) offers a large body of both linguistic and archaeological material which dissents from the above Bering Strait theory. She writes:

A study of aboriginal language distribution supports Knut Feldmark's hypothesis that the initial route of entry of peoples into the New World was along the Pacific Coast rather than through the interior ice-free corridor. The greatest diversification of aboriginal language as indicated by the number of language isolates and major subdivisions of language phyla, is observed on the Pacific Northwest Coast, in California, on the Northern Gulf of Mexico Coast, in Middle America, and in South America. Following a conventional principle of historical linguistics, it is assumed that the development of language diversification indicates a time depth of at least 35,000 years of human occupation of most of the Americas.

While the linguistic material is too specialised to be discussed here, there exists much independent archaeological evidence in support of Gruhn's thesis. Among others Gruhn concludes that 'the coastal entry model proposes that Pleistocene settlers were able to penetrate the New World initially following the Pacific Coast and exploiting littoral resources with a simple technology. The direct archaeological evidence – Pleistocene human occupation sites – indicates that the earliest people to settle in the Americas had a level of flaked stone technology more like the Lower Paleolithic than the Upper Paleolithic.'

However, Gruhn's further suggestion that all cultural developments in America that followed resulted from indigenous cultural evolution is mere conjecture and cannot be factually substantiated.

When I submitted Gruhn's article to George Carter for comment, he replied that he did not dispute the main body of her evidence, but that much of it needs amplification. Carter commented:

Gruhn is generally right but she ignores the highly probable complication of the language picture caused by Trans-Oceanic migration. Maya is Hamitic. There is Japanese language in Michoacan (Tarascan), Indo-European in

California. So some of the isolates are ancient, some recent. The picture is more complex than Gruhn sees it.

Carter also mentioned to me some of the results of a summit conference '*On the Peopling of America*', held at the University of Maine (May, 1989). Extracts of the conference proceedings mentioned by Carter are as follows:

> Niede Guidon presented a site in Brazil that she thinks bottoms out at 55,000 BP. De-Lumly of France was working with Maria Beltrano at another cave site and the date for that is said to be 300,000 BP (uranium dated on bone, by James Bischof at the USGS offices in California).

Carter notes:

> I discussed this with Bischof and he said the bone dates were very poor but that he made so many runs that he thought that he finally had reliable dates. So I just have to reiterate: Brazil: 300,000 BP also 55,000 BP Mexico, near Mexico City 250,000 BP; the Calico Site in the Mojave Desert: 200,000 BP; and more recently and still most controversial a possible hacked mammoth skeleton in the Borrego Desert (on the edge of the Imperial Valley in California) 500,000 BP I also hear of a 800,000 year BP date...

Carter concludes:

> The Old Guard are still fighting to hold the line at 11,500 BC. The more progressive are pressing for maybe 30,000 BP to 40,000 BP.

I am submitting the above as a conference report, though not necessarily accepting them as a gospel truth.

So far no fossil evidence has been produced to support the proposition that *Homo erectus* could have entered the American continent preceding even his Asian presence.

Other data given by Carter are as follows:

Pre-Columbian sites with some of the earliest radiocarbon dates showing human occupation are: Peru: Picimachay, 20,250 +/– 320 years BP; Chile: Monte Verde, 33,370 +/– 530 years BP; Brazil: Petra Furada, 32,160 +/– 1,000 years BP; Mexico: El Cedral, 33,000 +/– 2,700 to 1,800 years BP; California: Calico locality in the Mojave Desert is claimed to show definite characteristics of human flaking. Bishop *et al.* (1981) obtained a uranium thorium date of *c.* 200,000 years BP, Mexico: at the Valesquilo locality, Steen-Macintyre (1981) postulated an age of lithic Material older than 250,000 years BP. (So far, the dates of 200,000 and 250,000 years BP need further corroboration.)

As observed in this treatise. *Homo erectus*, and later *Homo sapiens* had already penetrated to very remote corners of the globe hundreds of thousands of years ago and reached these new places seemingly against all odds. This being proof that early humans had the ability and skill to cross large expanses of water. This proposition is incontestable even if we cannot now trace the routes these early humans took, or the means they used. Therefore, the early transmission of culture (i.e. certain lithic propensities) and its spread into the Americas must be held equally possible.

There is now evidence that Australia was reached by migrants arriving from South East Asia as early as 60,000 years ago. In this case, the respective fossils were dated by the electroscopic resonance method, Delson, citing Thorne: 1984 writes that, as this would have involved the crossing of large areas of water, the pertinent question raised by Thorne was that: 'If people have moved across ocean-gaps of a substantial nature in East and South East Asia at such remote times, may this not also be the mechanism which involved the process leading to the occupation of the Americas?' Although, it must be added, that when the aborigines crossed to Australia, the water gap was much narrower than it is today.

19

The Findings of R. A. Jairazbhoy

Jairazbhoy's work presents one of the most prolific sources for contacts between Pre-Columbian America and the Old World, based specifically on the discovery of Egyptian traits in the New World. The accumulation of his highly condensed material in three volumes (1974, 1976, and 1981) provides an extensive coverage linking the work of previous researchers with his own unique discoveries. To show the scope of his vision I am submitting a number of his illustrated examples. His more extensive list of 440 detailed items is too voluminous to be included in this book, and must be accepted on trust.

As to the presence of hieroglyphs in Mexico, the Americanist Jacques Soustelle in his book *The Olmecs* (1980), raises the objection that had the Egyptians visited Mexico they would undoubtedly have left hieroglyphic inscriptions. So far Jairazbhoy has discovered four hieroglyphs in Mexico (albeit a comparatively small number of Olmec sites have so far been excavated). Jairazbhoy's examples include the signs shown on page 124.

A vast amount of collateral material relating to every area of life links Pharaonic Egypt and the Middle East with Mexican Olmec and South America, and also of the later periods with the Far East. It should also be borne in mind that there are numerous examples which are in the nature of literary

'sky'   ;   'to give'   ;   'life' ('ankh' sign)   ;   'underworld'

The sign for *SKY* (1974: 50)

The Egyptian sky goddess Nut supporting the hieroglyphic sign for 'sky'.

Olmec Atlantis from Potrero Nuevo supporting a hieroglyph identified as the Egyptian sign for 'sky' (repeated four times).

The sign for *LIFE* (ank) (1981: 10)

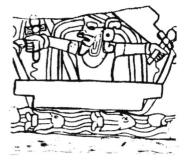

At the early Mexican site of Izapa on the Pacific coast is carved this relief of a bearded man seated in a boat. In each outstretched hand he holds what Norman describes as 'sceptre-like objects that do not resemble any other known Izapa or Mesoamerican implements, but do coincidentally resemble the Egyptian ankh sceptre, a symbol of life and prosperity'.

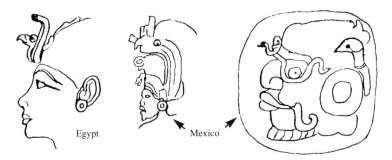

The uraeus serpent in Egypt and Mexico (1976: 53a, 53b, 53c)

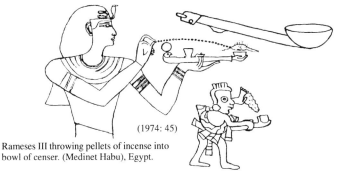

(1974: 45)

Rameses III throwing pellets of incense into bowl of censer. (Medinet Habu), Egypt.

Mexican priest, Codex Selden, throwing incense into bowl of censer.

Egyptian libation scene with Pharaoh purified by the gods Thoth and Horus. Notice crossed streams.

Scene from Mexican Codex with crossed streams of libation poured by two underworld gods over another.

(1974: 18/19)

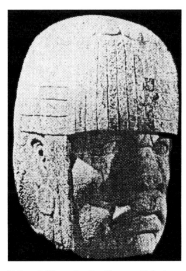 

Colossal Olmec head with negroid features now in Jalapa Museum. Observe helmet with linear markings (Mexico).

Egyptian heads from Tanis of Ramessid period wearing leather helmets. They portray traditional enemies including negroes (Montet).

This double jar with bridge handle from Peru has a close counterpart from Cyprus of the Bronze Age. Not only in that they are connected by a bridge handle, but they have the same general shape and proportions, and the only difference is the whistling monkey.

(1981: 38)

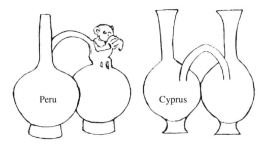

Peru

Cyprus

The hieroglyphic sign for 'underworld' (1974:81) in Mexico

Relief from Izapa with crowned human headed bird flying out of an opening resembling the Egyptian hieroglyphic sign for 'Underworld'.

While the *ka* appears in Eastern Mexico the Egyptian *ba* occurs on the West Coast. On a relief at Izapa a human-headed bird flies up out of an opening (*Norman, V. Garth: 1973, pl. 58*), just as in an Egyptian tomb painting a ba bird representing the soul of a deceased flies up out of a hole in the tomb (*Posener, G.: 1962, p. 267*). Few would doubt that the same conception is implied in both.

Egyptian papyrus painting with Ba bird flying up out of a hole in the tomb. The Ba is the soul of the deceased, and has freedom to come and go.

The use of cramps to hold masonry blocks together: Egypt, Bolivia. A photograph of the roofing blocks at Medinat Habu, Egypt, taken in 1976 (though there are earlier New Kingdom examples), and a photograph of the masonry from the Tiuhanaco site of Puma Punku, Bolivia taken in 1979.

(1981: 67)

evidence. R. A. Jairazbhoy has compiled a list of over 440 striking correspondences to support his case.

In trying to evaluate the evidence Jairazbhoy presents in support of the origin and rise of American pre-Columbian, civilisation (of which the illustrations are just a small sample), he elucidates (1976:101): 'No one with an open mind can with reason deny that long sequences of precise parallels between the two cultures must inevitably be the result of direct contact. I have set out such long sequences of precise parallels between Ramessid Egypt and Olmec Mexico.'

Jairazbhoy points out that cultural correspondences between Ramessid Egypt and early-Olmec Mexico fall into the same time context of c. 1,200 BC. The physical preconditions of the area where the first American civilisation, (that of the Olmecs) became established c. 1,200 BC were far from favourable (Michael Coe, 1968). Coe (a leading authority on the Olmecs) writes that: 'the abruptness of its appearance at the hot coastal plain of Southern Vera Cruz, has no convincing explanation at the present moment.'

And Jacques Soustelle, another expert, writing more recently (1985) about Olmec origins, points out that 'There is no evidence of "formative" evolution, a gradual maturation over several centuries. This indeed, constitutes the very head of the Olmec mystery … the astonishing spectacle of a civilisation that gives the impression of suddenly springing up in its originality from an undifferentiated background of peasant culture.'

According to Jairazbhoy, all this points to a major intrusion of trans-oceanic civilisers with an Egyptian cultural background.

20

The Views of Joseph Needham and Gwei-Djen

In contrast to Jairazbhoy who deals with both trans-Atlantic and trans-Pacific intrusions into the New World, the above authors deal mainly with the Pacific side (*Trans-Pacific Echoes and Resonances*, 1985). Doing so, they present us with a great number of cultural parallels concluding that they can only have arisen from direct contacts between the two hemispheres. Yet contrary to what their insistence on such numerous physical contacts suggests, they contend that these influences played no decisive part in the formation of the early American civilisations which they tend to ascribe to local origins. However, evaluating their findings fails to produce the criteria which would lend credence to such locally produced isolated beginnings. Lacking any material evidence to support their isolationist viewpoint, they turn as a last resort to Glyn Daniels's evolutionary solution, a surmise which has been dealt with in more detail in Part Two.

Furthermore in their theoretical considerations (p. 14), they quote Sir James Frazer (1966:459) as saying that: 'The independent evolutionists have to try and prove a negative case which is always difficult and only rarely do they offer explanations for the existence of parallels, though logically speaking such a burden is probably as incumbent on them as it is on the diffusionists.'

A more rational evaluation of the evidence suggests that Needham's and Gwei-Djen's examples lend strong additional support to the Old World origins of American civilisations, both in their formative and subsequent periods, although it must be admitted that at the present stage of research there is no conclusive proof for this. A selection of examples drawn from the large dossier cited by Needham and Gwei-Djen (1985) is given below.

About jade. Jade was equally treasured by the Aztecs, the Maya and the Chinese. While this alone is insufficient to postulate diffusion, the fact that both, the Maya and Chinese placed jade beads in the mouths of their dead, sometimes coloured with a life-giving red, must be accepted as beyond coincidence. An equally strong example is seen in the similarities between the Meso-American game of 'patolli' and the similar Asian (Indian) game 'pachisi'. E. B. Taylor, as far back as 1879 concluded, after making his theoretical definition, that trans-Pacific contact alone would explain the connection between the two games.

Mythology and cosmology offers a number of bizarre similarities common to Asia and America. Hatt (1953), is cited for details on the earth-diver or diving-god, the thunder-bird, the sky's up and down movement, magical shamanistic flight, the one-sided man, the swan-girl, the Orpheus motif, the stealing of food plants from the sky or their origin from a sacrificed child, the corn-mother, and many more. Every sinologist who visits Mexico is amazed that the Aztecs should have had the idea that there was a rabbit in the moon, since it is so characteristically Chinese (p. 30). Another curious similarity on both continents is the belief that the heavens are round and the earth is square, symbolically expressed on both sides of the Pacific in a circle and a rectangle (p. 35).

In technology, common elements are the spear-thrower and the balanced shoulder pole (according to Jairazbhoy, this has an Egyptian parallel in the coolie-yoke carried by farmers as shown in tomb paintings), Beirne (1971) in his study of axe and adze haftings selected five out of seventeen types as strongly suggestive of trans-Pacific diffusion (p. 40). Other features intrusive from Asia to America are plate armour, conch-shell trumpets, and the blowgun, the latter has been convincingly documented by Jett (1970). Alexander von Humboldt – as early as 1802 – noted the similarity of suspension bridges between those at Ponipe in the Americas and those he had encountered in Bhutan and western China. As for sluices, irrigation channels and dams in the New World, Enok has described likely correspondence with the Old World at length (1912:40/47).

The sailing rafts of the Peruvian and Ecuadorian coasts and those of South China, Taiwan and Vietnam, are described as 'blatantly similar'. The authors stress that it is almost quixotic to deny any connection between them (Doran, Meggers *et al.* 1971 and 1975). Furthermore if the similarities between the Valdivia pottery of Ecuador and the Jomon pottery of Japan is provable beyond doubt, it would mean that Pacific voyages took place as early as the beginning of the third millennium BC (Meggers, Evans, Estrada, 1965). Many scholars, such as Gardini, Tolstoy, and Kelly, have been inclined to accept the close relationship and influence of the Valdivia Jomon finds.

According to the authors, Alfonso Caso (1964) refused to accept the nautical evidence of the possibilities of contact by sea between Asia and pre-Columbian America. He considered that even if detailed similarities should be proved between Meso-American calendrical and cosmological formulations and those of the Old World, they would never be sufficiently conclusive to prove actual contact. The authors comment that with all due respect for Caso, one must reject such an extreme isolationist position. They agree, however, with Caso that no specifically identifiable pre-Columbian objects have so far turned up on the American continent.

While this point needs further investigation, one can perhaps argue that the presence of alien scripts in Pre-Columbian America could be considered as an equivalent of foreign objects. This the more so as none of these fully fledged foreign scripts can under any circumstance be ascribed to independent origins. In this respect the large amount of pre-Columbian

foreign scripts intrusive into America, discovered and translated over the years, by epigrapher Barry Fell, though hotly
disputed, must be given serious attention.

Bark-cloth making. Tolstoy distinguished 121 traits of which 92
were shared by South East Asia and Meso-America. In Asia
origins go back to 2,000 BC, in America to 880 BC (Tolstoy,
1963), Kelley and Meggers stress that the similarities in techniques for making bark-cloth are inconceivable without admit-
ting trans-Pacific contact (pp. 48–52).

Ethno-botany, ethno-zoology. Here the authors regrettably de-
preciate the spadework of G. F. Carter, who made important
contributions in his *Plants Across the Pacific* (1953) and related
papers. Particular attention ought to have been given here to
his penetrating study 'A Hypothesis Suggesting a Single Origin
of Agriculture' (submitted to the Congress of Anthropological
and Ethnological Sciences, Chicago 1971). Furthermore, although the authors comment favourably on his (Carter's) treatise
of the domestic fowl ('Pre-Columbian Chicken in America',
1971), they throw doubt on some of his findings (pp. 60–63).

Conclusions. Needham and Gwei-Djen point out that in spite
of the great wealth of comparative material from Asia intrusive
into pre-Columbian America, nothing in their review deters
them from accepting what they call the 'formidable originality'
of the American cultures. They confirm a position already taken
in their introduction (p. 6), where they endorse adherence
to the views of Glyn Daniel who had insisted on the independent origins of the world's early civilisations in general and that
of the American pre-Columbian civilisations in particular (see
The First Civilisations, 1968). Remarkably though, the question
of how these isolated events of native origins could have
occurred is neither answered by Needham and Gwei-Djen, nor
by Daniel. Instead of citing supporting facts, Needham and
Gwei-Djen uncritically endorse their Cambridge colleague's
dubious theory of multilinear cultural evolution as having been
the catalyst of American civilisations. Glyn Daniel had written
(*The Idea of Prehistory* 1964:105), 'The origin of American
civilisation was in fact a tale of independent cultural evolution.'
Despite this theoretical aberration in the evaluation of their

evidence, the two authors Needham and Gwei-Djen must be praised for having gathered and added such a large dossier of proven cultural contacts between Pre-Columbian America and the Pacific hemisphere, to an already existing multitude.

21

The Contributions of Professor J. L. Sorenson

Throughout this work emphasis has been laid on the fact that without recourse to cultural evolution (a theory now found dubious) the independent and rather sudden emergence of pre-Columbian American civilisation cannot be rationally ex-plained. Apparently no other valid alternative explanation apart from historical contact has been offered up to now. One of the few Americanists who has recognised the importance of the evolutionary cultural tenet's possibly negative impact on cultural theory is Sorenson. In *Man Across the Sea* he compiled an impressive list of cultural parallels under the heading '*The Significance of an Apparent Relationship Between Ancient Near East and Mesoamerica*' *(1971:219). Among other things he states:*

> At no time in the modern era of anthropology has a serious attempt been made to relate the historical issue to theoretical currents in anthropology, other than incidentally to evolution.
> . . . Dissatisfaction with both the substance and form of evi-dence for Old and New World connections as certain diffusionists have presented it, has prompted me, over the last fifteen years, to accumulate my own corpus of evidence. My present professional position and interests ensure that so long as this corpus of data

remains only in my files, it will not be criticised, clarified or supplemented as it ought to be. In view of the significance of the issue involved, I am obliged to offer the material for use by scholars who may be in a position to consider it more fully.

> *... Unless further research disqualifies a great deal of the evidence presented here, one must conclude that a substantial number of the cultural features of much more than peripheral significance in Mesoamerican civilisation either originated or were at least present even earlier in the heart of the Old World oikoumene.* The exact medium of communication, the times, the route, and the specific content transmitted are unclear, just as details about Mesopotamian-Egyptian communication in the Proto-literate (see Frankfurt, 1954) remain hazy, yet *the fact of relationship is forced upon us by the circumstantial evidence* [my italics].

[Sorenson concludes] On the basis of the evidence presented, it is plausible and perhaps necessary, to interpret the rise of civilisation in Mesoamerica as significantly dependent upon communication from the heartland of Eurasia through intermediate steps that are unclear.

Now almost 20 years later, to crown his work, Sorenson has produced *A Comprehensive Annotated Bibliography* (1988), which he subtitles 'Study Aid'. It deals with Trans-oceanic cultural contacts between the Old and the New Worlds in Pre-Columbian times. This massive work is not only a veritable gold mine for knowledge-seeking Americanists but generally also for prehistorians the world over (the work has since been re-edited in book form).

In the process of editing, Sorenson and his assistants scanned thousands of books and periodicals, both in English and other languages. The work, stretching over several years, contains hundreds of abstracts.

On the question of whether the origin of pre-Columbian American civilisations was due to isolated indigenous origins, or due to trans-oceanic incursions, Sorenson remains largely non-committal. Although his sympathies appear to lean more towards diffusionism, he does not reject the possibility that convergence and independent parallel developments may have played some part. In his interim conclusions he states: 'What ought really to be addressed is the far more complex, and uncomfortable question, who crossed, where, when, and by what means, and precisely what did and did not result from such activity.'

My own judgement is that Sorenson's 'Study Aid' adds over-whelming evidence, to that already presented in my present work in favour of the world's close cultural inter-relationship. Although Sorenson grants to 'convergence' and 'independent parallel invention and development' a major role in culture growth? I have so far found scant evidence that these factors have played any decisive part in the worldwide shaping of cultural history particularly now that the idea of cultural evolution can be shown to be obsolete.

Among authors who, like Sorenson, grant to cultural diffusion a decisive part in the formation of pre-Columbian American Civilization but hesitate to discard the influence of 'convergence' and 'independent invention and development' as a factor in the process, are Alice Kehoe and David Kelley. On the other hand, George Carter, whose contribution to American anthropology is greatly underestimated, stands firmly by a prominently diffusionist solution, and so does R. A. Jairazbhoy.

John L. Sorenson commenting on Olmec origins writes (A Mesoamerican Chronology, April 1977. Katunob 9(4) February 1977: 41–55):

> One must be sure not to suppose that 'Olmec' means only the development along the Gulf-Coast typified by the sites of La Venta and San Lorenzo. Actually highland sites are contemporary or earlier but do not fit comfortably under the 'Olmec' rubric and are not particularly derived from the lowlands. I should say that Coe has not made a 'finding' but an assertion. And where you offer a date of 1,200 BC for Olmec beginnings, it ought actually to be no later

than 1,500 BC (the 1,200 is literal radiocarbon years), when corrected that comes out at least 1,500 BC, and this is strictly for the Gulf Coast manifestations.

22

African Presence in Early America:
BC18 and AD

In his essay of the above title, Ivan Van Sertima illuminates the
historical role African (specifically Negro People) played in the
opening up of early America. And although pre-Columbian
Egyptian intrusions cannot be strictly attributed to the Negro
section of humanity, one has to keep in mind that African
achievements have their biological and cultural origins in a
basically dark-skinned African rootstock.

In submitting his case van Sertima draws on a group of
collaborators, including Alexander von Wuthenau (unexpected
African faces in pre-Columbian America); Keith Jordan (evidence from physical anthropology); Joan Covey (evidence from
early maps); Beatrice Lumpkin (pyramids – American and
African).

The AD section deals mainly with a re-examination of Leo
Wiener's work and the Mandingo voyages, while the Asian Part
(Africans out of Asia) deals with the Harold Gladwin's thesis and
other African-Asian connections – dealt with, among others by
Runoko Rashidi, Legrand Clegg II, and Wayne B. Chandler.

What is remarkable, is, that most of van Sertima's co-authors
take recourse to the groundwork of Rafique Jairazbhoy, who
cites such striking Old World parallels as: monarchic traits; the

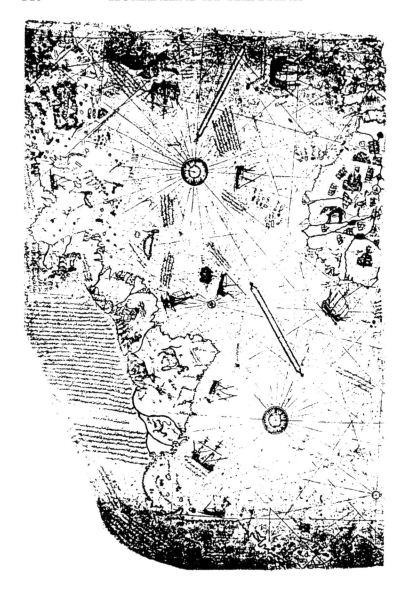

The Piri Re'is map, showing parts of South America

double crown; the royal flail; sacred boats; royal purple; artificial beards; feathered fans; ceremonial umbrellas; incense spoons; the four Babacs holding up the sky; the opening-of-the-mouth ceremony; human-headed bird figure emerging from tomb and many others. They all indicate Egypt-inspired intrusions into pre-Columbian America and even point to the possibility of direct Egypto-American contacts. Jairazbhoy's work is given specific prominence in van Sertima's own contribution entitled 'Egypt-Nubian Presence in Ancient Mexico'.

The most original contribution in a field never before treated in such detail is Joan Covey's article 'African Sea Kings in America? Evidence from Early Maps'. Covey presents clear evidence in the form of numerous ancient maps which predate 1,500 AD, recording that their authors could already measure longitude, an art subsequently lost, being only re-discovered 250 years later.

Furthermore, there are maps predating Columbus which show accurate features of American topography, the most famous being the Piri Re'is Map. This map of 1513, was only re-discovered in 1929. Covey quotes Hoye who wrote that: 'In 1955, a cartographer named M. Descombes announced the discovery of Ferdinand Magellan's own chart of his epochal circumnavigation of the world. No one had known it existed, but Descombes found it in the archives of Istambul. It contains information, which according to the history books, was not available in 1513, namely that South America and Africa are in correct relative longitude.'

How did Piri Re'is get such accurate data? Piri Re'is wrote that he used 20 source maps; among them eight dating from the time of Alexander the Great. The features shown on the Piri Re'is map (the Andes, the Atrao and Amazon rivers, and the island of Marajo) were obviously shown on the source maps which Piri Re'is consulted. Someone had not only known of South America by then but had mapped part of it at least hundreds of years before Columbus sailed west.

Joan Covey concludes that the large amount of evidence presented indicates that African sea kings went to America. 'It is not likely that we can ever prove that they used maps of

America on their expeditions. Maps of America may have been in circulation in the Nile Valley area (the Middle East) at that time. Future finds may shed more light on this.'

23

The Astonishing Dossier of Sir James Frazer

The learned reader may express surprise and be even stunned by finding the work of James Frazer revived here, 100 years after it was published (1896). Yet the world-spanning material collected by Frazer, as the record will show, is as topical now as it was then, having no equal in anthropological literature.

The only parallel I can draw to his work (not his cultural theories), i.e. his collected ethnological material, is with Charles Darwin's epoch-making *The Origin of Species*. Despite being published about 140 years ago (1859), Darwin's work is as topical today as it was then. Thus, to emphasise the global significance of cultural parallels I am submitting an extensive catalogue of worldwide cultural parallels of past and present, from Frazer's *The Golden Bough* (1960 edition).

Frazer's *The Golden Bough* is a unique collection of cultural parallels, drawn from an amplitude of societies worldwide. On this basis, Frazer constructed universal theories of human society and institutions. Seemingly, Frazer's deductions, while supportive of the independent invention and development theory in culture, rest mainly on assumptions of the psychic unity of all mankind, as formulated by Bastian. In contrast to this assumption, the use here and interpretation of Frazer's formidable mass of material can now be seen (disregarding Frazer's psychological interpretation) as providing a mass of evidence for the universal inter-relationships of most of the world's cultures and indicates cultural spread and diffusion from common sources of origin.

What becomes also evident from Frazer's examples is that practically all fall into a relatively narrow time-context, dating from the emergence of the formative stages of the earliest civilisations (*c.* 4,000 years BC) to modern times. On closer examination, Frazer's examples relate principally to those culturally more complex social groupings (hunter-gatherers excluded) which are associated with agriculture, stock rearing and civilisation. As previously noted, these groups existed during less than one quarter of one per cent of hominid history, which totals *c.* four million years; and during less than two per cent of *Homo sapiens's* cultural history, which totals *c.* 300,000 years. In contrast, food-gatherers lived for untold generations within small family groups or bands, and showed no tendency towards forming more complex societies. Although their history is about one thousand times longer. This latter enormous timespan, is almost void of any of the social customs and institutions which Frazer presents in such amplitude in *The Golden Bough*.

The Ritual of Osiris and its Repercussions

In studying Frazer's work one is induced to conclude that Osiris's role in Egyptian history (both as king and as god) had far-reaching consequences not only for Egypt, but the ideational history of the world. Osiris emerges as a veritable fountainhead of numerous worldwide cultural parallels, whose origins can be reconstructed from historical sources and mythical accounts.

As to historical sources Emery (1984:122/123), who spent much of his life excavating in Egypt, wrote:

> Until recently it was questioned whether the worship of Osiris had developed in archaic times. The discovery at Helwan (situated south of Cairo), of the 'Dad symbol', of

the god Osiris and the girdle-ties of his female counterpart Isis (both of First Dynasty date), show that the cult
which was to remain the most favoured of the masses of
Egypt throughout the long history, was already in existence then. The cult, although reminiscent of nature worship, was primarily the worship of dead kingship, and the
myth of Osiris seems to be an echo of long forgotten
events which actually took place. These events were Perhaps originally unconnected and belonged to different
periods, but they were later welded together into a moral
story of the struggle between good and evil. The myth of
the treacherous murder of the good king Osiris by his
brother Set, and the vengeance and re-establishment of
beneficent rule by Osiris's son Horus, who founded the
line of demi-gods from whom the pharaohs were descended, all suggests episodes perhaps connected with the
prehistoric struggles between the dynastic peoples and the
indigenous of the Nile valley. [Emery concludes] that all
these stories were perhaps based on actual historical
events.

Concerning mythical accounts about Osiris, several versions
exist, with one of the earliest authored by the Greek historian
Plutarch (*c.* AD 100). Some modern authors consider Plutarch
outdated, pointing out that modern translations of pyramid
texts differ substantially. In this respect, particular reference is
made to J. G. Griffith's book, *The Origin of Osiris* (1966). Yet on
studying Griffith one finds that his account differs only in
detail, not in substance, from Plutarch. As a matter of fact,
Griffith himself frequently refers to Plutarch (see Griffith, pp.
13, 14, 17, 18, 37, 61, 71, 134 and 174). As to the alleged
outdatedness of Plutarch's account, as compared to modern
texts, one can counter that Plutarch may have had access to
sources which in the intervening 2,000 years were lost.
    For the historian the relevance of Osiris is not only his
unique origin but more so the subsequent worldwide similarities
in rituals and customs which resulted from the Osiris cult.
According to Frazer (p. 477) Plutarch's version forms the

only connected and reasonably documented extant narrative. Margaret Murray (the well-known Egyptologist, 1962:127) agrees with this assessment. She writes: 'Though Plutarch (in 'De L'side de Osiride') is a late writer, his account of Osiris can be checked by the records of Osiris-worship, and is proved to be substantially correct.' Part of the Osiris story relevant to this paper is that: Osiris was the offspring of an intrigue between the earth-god Seb (Keb or Geb) with the sky goddess Nut, resulting in the birth of Osiris. But Osiris was not Nut's only child, others being the elder Horus, Set, Isis and Nephtys, all of whom became deities. In subsequent incestuous unions, Set married his sister Nephtys and Osiris his sister Isis. J. G. Griffith writes (1966:132) that Osiris was believed to have married a goddess who was also his sister. This is clear from the earliest accounts – Text 37 Pyr 172a (W) 'Isis, this Osiris is thy brother.' In the same spell, Osiris is referred to as the brother of Nephtys and Set, and Thoth as the son of Atum, Shu, Pefnut, Geb, Nut and as the father of Horus; and Isis as the sister of Osiris.

Plutarch records: 'As king on earth Osiris reclaimed the Egyptians from savagery [i.e. introducing civilisation, G.K.] gave them laws and taught them to worship gods.' He also introduced the cultivation of wheat and barley, fruit trees and grapevines, and subsequently traversed the world, diffusing the blessings of agriculture and civilisation everywhere. On his return his brother Set with fellow conspirators tricked him into lying down in a coffer and then fastened the lid, causing his death. The coffer was thrown into the Nile, was swept out to sea, and eventually landed at Byblos on the coast of Syria. A distressed Isis (his sister and wife) searched and eventually found the coffer with the dead Osiris still inside and carried it back to Buto in the Nile Delta.

One day when Isis was out looking for her son young Horus, Set found the coffer while on a boar hunt. He removed the corpse of Osiris and cut it into 14 pieces which he scattered all about. The historian Diodorus Siculus, claims that Isis recovered all the pieces except the genitals. Wishing that her husband's real grave should remain unknown, Isis moulded around each of the parts an image of Osiris and asked her priests to bury them all over the country. A long inscription in the temple of Denderah has preserved a list of the god's burial sites. The inscription at Denderah (40 miles north of Thebes) and other texts mention the parts of Osiris treasured as relics in each of the sanctuaries. The heart was located at Athribis, his backbone at Busiris, his neck at Letopolis, and his head at Memphis. But duplicates abounded. For example, another Osiris head was claimed for Abydos and numerous of his legs elsewhere. A long inscription in the temple of Denderah has preserved a list of the god's burial sites (Fraser 1960:481/482).

It is further recorded that when Isis had found Osiris's corpse, she and her sister Nephtys broke out in lamentations. Feeling pity for their sorrow, the sun god Ra sent down from heaven the jackal-headed god Anubis who, with the aid of Isis, Nephtys, Thoth and Horus, pieced together the broken body of the murdered Osiris and swathed it in linen bandages, while enacting the prescribed funeral rites. According to Griffith (1966), it is clear from the references that Osiris is the prototype, even if one that reflects ultimately the king's condition. He is the god who was treated in this way after death and the deities who did it for him are Horus, Geb, Isis and Nephtys. The text reads (Pyr 1683 b–1685 a): 'I make thee live, I gather for thee thy bones, I collect for thee thy mutilated parts: indeed I am Horus who saves his father', and, (Pyr 1981 a–c):

> Thou [Osiris King] art washed by Isis, Nephtys has cleaned thee, the two great and mighty sisters gather together thy flesh and fasten thy limbs and make the two eyes appear in thyne head; and [Pyr 2008 a–b, of 1292 c] . . . Thy bones are collected for thee, thy dust is shaken off thee, thy bones are loosed for thee. Then Isis fanned her wings over the body, whereupon Osiris came to life again and henceforth reigned as king of the Nether-world.

All the various events surrounding the myth of Osiris, from his murder, his dismemberment, the scattering of the pieces and their later reassembly to his eventual resurrection, had a great impact not only on Egyptian imagination and beliefs but led to worldwide repercussions, Osiris's life and death was annually re-enacted in many parts of Egypt; while the legend that Osiris brought agriculture into the world, led to the belief that it was he who dispensed fertility to the land and caused the grain to grow. The fact that the various parts of his body had been buried all over the country led further to the belief that in order to restore fertility to the soil the act of identifying the annual revival of the vegetation with the resurrection of Osiris was essential.

In a chamber dedicated to Osiris in the great temple of Isis at Philae there is a bas-relief which shows the dead body of Osiris

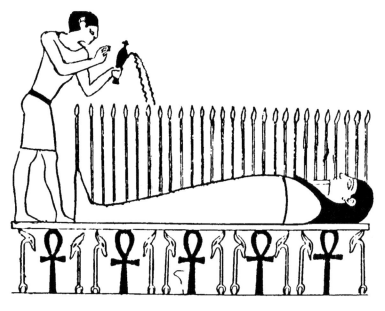

Here it is shown that Osiris the corn god produced the corn from his body. In other words, he gave his own body to feed the people; he dies that they may live. Frazer concludes (p. 497), 'The foregoing survey of the myth and ritual of Osiris may suffice to prove that in one of his aspects the god was the personification of corn which may be said to die and come to life again every year. Also, the story that Osiris's mangled remains were scattered and buried in different places indicates a symbolic expression of both the sowing and the winnowing of the grain.'

with corn stalks sprouting from it, while the priest waters the stalks from a pitcher.

In Egypt, the Nile was the source of all life and the regular arrival of the Nile flood was an annual occasion for religious ceremonies. Coincidentally, the star Sirius, the brightest of all stars, appears at dawn in the east as the Nile begins to rise. This was interpreted as signifying that heavenly influences regulated earthly events. In the process of such symbolism, Isis, the wife of Osiris, became identified with the star Sirius (the Egyptian Sothis) and the tears she shed lamenting Osiris's death were believed to cause the Nile's rise and inundation.

Another version to account for the Nile's regular flood was directly attributed to Osiris. Erman (1927:7) in a translation of an 18th dynasty hieroglyphic text records about Osiris: 'The Nile issueth ... from the sweat of thine hands. Thou spittest out the air which is in thy throat into men's noses ... trees and herbs, reeds and ... barley, wheat and fruit-trees ... Thou art the father and mother of mankind. They live on the breath, and they eat off the flesh of thy body. Primordial God is thy name.'

In another text from the time of Rameses II, quoted by Erman (1927:80), Osiris is directly identified with the Nile, i.e. 'Thy nature Osiris, is more secret ... Thou appearest in order to dispel darkness ... Verily thou art the Nile ... men and gods live by the moisture which comes from thee.'

The Egyptian harvest falls in the months of March/April/ May. It is recorded that the ancient farmers, while feeling secret joy at the bounty of the harvest, also showed dejection at being forced to sever the body of the corn god with the sickle and trample it under the hoofs of animals on the threshing floor. In atonement for such sacrilege the corn reapers beat their breasts and lamented over the first sheaf cut.

According to Frazer, similar laments were chanted by corn reapers in Phoenicia and other parts of Asia, mourning the corn god (whom they called Maneros) killed by their sickles; in Egypt, the slain deity was Osiris.

Plutrach, quoted by Frazer, records that Osiris was buried on the seventeenth day of the month of Athyr and that the Egyptians accordingly observed mourning rites from that date onwards. These ceremonies were portrayed in drama: first, recalling the search for the dead Osiris, and second, its joyful discovery, followed by the resurrection of the dead god in the image of vegetable mould and spices. The ceremonies varied in different places, but all set forth the triple aspect of Osiris; as dead, dismembered, and finally reconstituted by the re-assembly of his scattered limbs.

Human Sacrifice to Promote Fertility

Frazer shows that parts of the Osiris myth are reminiscent of worldwide practices of sacrificing humans in the image of the corn-spirit (identified with Osiris), distributing their flesh and scattering their ashes over the fields to fertilize them. According to the historian Manetho, the ancient Egyptians used to burn red-headed men and scatter their ashes with winnowing fans and it is highly significant that this sacrifice was offered by the king at the burial sites of Osiris. Frazer (p. 498) also quotes examples from modern Europe where the figure of death is sometimes torn to pieces and fragments buried in the ground to ensure a good crop.

Osiris and his brother Set are also recorded as having been symbolically torn to pieces after a reign of 18 days an event commemorated by the annual festival of the same length. In a Greek legend, Pentheus, king of Thebes, and Lycurgus, king of the Thracian Edonians were rent to pieces for offending Dyonisos the wine god; the one by frenzied bachanales, the other by horses; so that the ground which had ceased to be fruitful might regain its fertility. In northern Europe the body of the Norwegian king, Halfdan the Black, was cut up and buried in different parts of the kingdom for the sake of ensuring the fruitfulness of the earth. In ancient Mexico, new-born babies were sacrificed when the maize was sown, older children when it had sprouted, and so on, until it was fully ripened when they sacrificed old men. An alternative name for

Osiris was the 'crop' or 'harvest'; with the ancients sometimes depicting him as the personification of the corn.

Other indications of the worldwide spread of Osiris-associated rituals abound. The Indians of Guayaquil in Ecuador used to sacrifice human blood and the hearts of men when they sowed their fields. The people of Canar (now Cuenca) in Ecuador sacrificed 100 children annually at the harvest. The Pawnees, a North American Indian tribe, sacrificed a human victim in the spring when they sowed their fields thinking that failing to do so would cause total failure of their crops of maize, beans and pumpkins. One recorded event, in 1837 or 1838, was the selection of a 14 or 15 year old Sioux girl for sacrifice. After six months of preparation, her body, having been painted half red, half black, was attached to a gibbet and roasted for some time over a slow fire, then shot dead with arrows. A medicine man next tore out her heart and devoured it, and while the remainder of her was still warm it was separated in small pieces from the bones, placed in little baskets and removed to a corn field where drops of blood were squeezed on the newly deposited corn in the hope of obtaining a bountiful crop.

The Marianos, a Bechuana tribe in Africa chose as a victim a short, stout man to be killed amongst the wheat to serve as 'seed'. Some parts of the body were then burned and scattered over the ground as fertiliser, and the rest eaten. The Bagobos of Mindanao, before they sowed their rice, killed a slave who was then hewn to pieces. On another Philippine island, Luzon, the natives of Bontoc were passionate head-hunters during rice-planting and reaping seasons. To assure good crops, every field needed at least one human head. Another Luzon tribe, the Apoyaos had similar customs.

Among the Indian Lhota Naga, it once was a common practice to chop off the heads, hands and feet of strangers and stick up these extremities in their fields to ensure good crops. At least once they flayed a boy alive, carved him in pieces, and distributed the flesh among the villagers to avert bad luck and ensure plentiful crops. The Gonda of India kidnapped Brahman boys to be sacrificed at sowing and reaping times. One boy was slain with a poisoned arrow and his blood then sprinkled over rice crops and ploughed fields while his flesh was eaten. The Oraons of Chota Nagpur offered human sacrifices to a goddess called Anna Kuari; the victim's throat was cut and the upper part of the ring-finger and of the nose were carried away. The goddess was then assumed to be residing in the house of the sacrificer and henceforth his fields would yield a double harvest. In China Kimedy, Bengal, a victim was dragged along be field, while the crowd, avoiding his lead and intestines, hacked away his flesh, while life still remained. In another district the victim was slowly roasted to death and the following day the body was cut to pieces; divided among heads of families, and carried to the stream which watered their fields. At a later stage, animals were substituted for humans. The populace would fall upon the tightly bound animal, hacking it to shreds and as soon as anyone had secured a piece, they would rush off and bury it in their fields.

According to Frazer, evidence from Egypt records that the slain corn spirit (the dead Osiris), was represented by a human victim, whom the reapers slew on the harvest field and later mourned in a dirge. The legend of Busiris (which besides being a town in the Nile delta was also the name of the local king) seems to have preserved a reminiscence of human sacrifice in the worship of Osiris. The king of Busiris, in Greek times, sacrificed strangers on the altar of Zeus. The origin of the custom relates to a famine which had afflicted Egypt for nine years. A seer informed Busiris that the famine would end if humans were annually sacrificed to Zeus. Frazer (p. 583) points to the almost identical behaviour and rituals of the Pawnee Indians of North America, believing likewise that the omission of human sacrifice would be followed by a total failure of their crops.

The place Busiris was in actual fact the ancient city Per Asar (House of Osiris), named after him because it was claimed to be his original home and contained his grave. Some authorities believe that from this source the custom of human sacrifice spread to other parts of Egypt.

Since there is no other adequate explanation for the global practice of human sacrifice to promote fertility, derivation from a common source – i.e. the Osiris cult – becomes a strong possibility and, if so, Egypt has to be the catalyst because it was attested there earliest and survived there longest.

Eating the God and Cannibalism

As a god of vegetation, writes Frazer (pp. 501, 502), Osiris was naturally considered as a god of creative energy. Even in death his generative virtue was not extinct but only suspended. Among the hymns addressed to Osiris is: 'Thou art the father and mother of mankind, they live by thy breath, they subsist on the flesh of thy body' (Erman, 1927:7).

This points directly to the Christian doctrine of trans-substantiation, wherein the whole substance of the bread and wine is symbolically converted into Christ's body and blood, an act which decidedly carries the stigma of cannibalism. Frazer (p. 576) also tells us of a Pawnee chief (North American Indian) who devoured the heart of a Sioux girl while the Marimos and Gonds ate their victim's flesh. He concludes that if, as we suppose, the victim was regarded as divine, it follows that in eating his flesh, his worshippers believed themselves to be partaking of the body of their god.

Among the Aztecs the custom of eating the body sacramentally in the form of bread was practised before the discovery and conquest of Mexico by the Spaniards, and was a bi-annual practice. Here the Christian doctrine of trans-substantiation has its pagan precedent. The ancient Mexicans believed that by consecrating bread their priests could change it into the very body of their god, so that all who thereupon partook entered into a mystic communion with the deity by receiving a portion of his divine substance into themselves.

The same doctrine, according to Frazer, was also familiar to the Aryans of ancient India, long before the spread, and even before the rise, of Christianity.

Similar customs prevailed in Europe. In Wermland, Sweden, the corn spirit was believed to reside in the last sheaf harvested; and to eat a loaf from it was to eat the corn spirit itself. Similarly at Las Palisse, France, representing the corn spirit, dough was moulded in the form of a man and this was baked and later eaten. Similar customs are also reported from Lithuania and Estonia. Frazer also records that in one part of Yorkshire it is still customary for the clergyman to cut the first corn then used to make the Christian communion bread. The Ainu of Japan make millet cakes for the same purpose and after worshiping them, they eat them. Frazer remarks that no doubt the cereal offering is regarded as a tribute paid to the god, but the god is no other than the seed itself; and it is only a god in so far as it is beneficial to the human body. Similar practices occur in many other tribal societies too numerous to mention.

The Transfer of Physical Attributes (trans-substantiation)

A variation of the motif of eating the god in order to become endowed with his spirit and powers, devolved into numerous worldwide practices. Among the examples given by Frazer (pp. 651–54) are these: Kansas Indians in North America as well as men of the Buru and Aru Islands of the East Indies would eat the flesh of dogs to acquire their boldness and nimbleness before going to war. The Zulu chief Matuana is reputed to have devoured the gall bladders of 30 enemy chiefs whose people he had destroyed. And with variation, identical practices were carried out in New Zealand. The global spread of such practices include: Papuans in New Guinea; natives of Northern Australia; natives of Korea; a Norse king; natives in Darfur of Central Africa; the Basutos of Southern Africa; the Ashantees of the Gold Coast (present-day Ghana); the Naura Indians of New Granada; the Tolalaki of Central Celebes; the Italones of the Philippines and natives of New South Wales in Australia.

Adonis, Attis and Tammuz

A further perpetuation of Osiris-inspired customs can be found in the myths of Adonis, Tammuz and Attis. Frazer comments (p. 428):

> Nowhere, apparently, have these rites been more widely and solemnly celebrated than in the lands which border the eastern Mediterranean. Under the names of Osiris, Tammuz, Adonis and Attis, the peoples of Egypt and Western Asia represented the yearly decay and revival of life, especially of vegetable life, which they Personified as a god who died and rose again from the dead. In name and detail the rites varied from place to place: in substance they were the same, namely the supposed death and resurrection of this oriental deity, a god of many names but essentially one nature.

While searching for the origin of such ideas and customs, Frazer remained fettered to the evolutionary, psychological conception of independent origins (Bastian's psychic unity of all mankind), which conjectures that all such and similar ideas arose spontaneously in different geographical areas. Without this strong preconception, it is reasonable to presume that Frazer might quite likely have concluded that some of these deities sprung originally from the Osiris myth, even if only for reasons of chronological precedence. And how could it have been otherwise? Frazer himself, under the heading 'Adonis in Cyprus' (p. 434), points out that Phoenician kings ruled in Cyprus until the time of Alexander the Great: 'Naturally they brought their gods with them from the mother-land. They worshiped Baal of the Lebanon, who may well have been Adonis, and at Amathus on the south coast they instituted the rites of Adonis and Aphrodite (or rather Astarte). Here as in Byblos, the rites resembled the Egyptian worship of Osiris so closely that some people identified the Adonis of Amathus with Osiris.'

According to Frazer, Attis was to Phrygia what Adonis was to Syria. Like Adonis, he appears to have been a god of vegetation, and his death and resurrection were annually mourned and rejoiced in the spring festival. The legends and rites of the two

gods were so much alike that the ancients themselves sometimes saw them as one. Attis was said to have been a fair young
shepherd or herdsman beloved by Cybele, the great Asian
goddess of fertility, who had her chief home in Phrygia. While
some held that Attis was her son; his birth, like that of many
other heroes, was said to have been miraculous, with his mother
being the virgin Nana.

This and similar other examples of virgin-birth can reasonably be presumed to be antecedents from which the virgin birth
of Jesus Christ could have originated.

The Corn mother and Corn maiden

History shows that the cult of the corn Mother and its many
variations (originating with Isis) became just as widespread as
that of the corn god (Osiris). And just as Osiris became the
corn spirit or corn god, his wife (and sister) Isis, became the
corn goddess. The Greek counterpart of Isis was Demeter who is
described as: 'she who has given birth to the fruits of the earth
and mother of the ears of corn.' In other words, the corn
mother Isis, or Demeter, is often represented by Greek and
Roman artists with ears of corn on her head and in her hand.
Furthermore, Demeter with her child Persephone are occasionally called 'corn mother' and 'corn child' (Frazer, p. 504).

Just as the ancients had their corn mother, Europe had its
wheat mother and barley mother, and Asia its rice mother. For
Europe, analogies have been collected in great abundance by
W. Manhardt. In Germany, when the corn waves in the wind in
springtime, the peasants say: 'There comes the corn mother; or
the corn mother is running over the field'. Children wishing to
enter the fields to collect flowers are warned that the corn
mother is sitting in the corn and will catch them. She is also,
according to the crop, called the rye mother, the pea mother,
or even the flax mother. In a village in Styria (Austria), it
is said that the a corn mother, in the shape of a female puppet
made out of the last sheaf of corn harvested and dressed in
white, may be seen at midnight in the cornfields, fertilising

them as she passes; but if she is angry with a farmer his corn will wither.

In Danzig (the Polish Gdansk), as well as in Holstein Germany, he who cuts the last ears of corn fashions a doll, which is called 'corn mother' or 'old woman'. In France, near Auxerre, the last sheaf goes by the name of 'mother of the wheat', barley, rye or oats. In the province of Osnabrueck, in Hanover, it is called 'harvest mother'. In Scotland, when the last corn was cut after Hallowmass, the female figure then produced was called 'Carlin' – the old woman', while amongst the Highlanders it was known as 'the old wife (cailleach)'. In Poland the last sheaf is commonly called 'the baba', that is, 'the old woman'. In Lithuania, the last sheaf is named 'Boba'. In Russia, it is often dressed as a woman, and in Bulgaria the last sheaf becomes a doll called 'corn queen' or 'corn mother'. Near Neisse in Silesia, an oats king and oats queen, seated in bridal array, are drawn by oxen through the village.

Eight thousand miles away, in Peru, the maize mother was represented by a puppet made of stalks of maize dressed in full female attire, and the Peruvian Indians believed that 'as mother' it had the power of producing and giving birth to large quantities of maize. The Menangkabquers of Sumatra, on the opposite side of the globe (like the Javanese), think that the rice is under the special guardianship of a female spirit called Saning-Sari, being represented by certain stalks of grains called 'Indoeapadi', literally meaning 'mother of rice'. Among the Tomori of Celebes, after the reaping is completed, the mother of rice, having previously been formed of stalks of rice tied together, is cut down and carried into a rice barn, and the other sheaves are all piled on top of it. In the Malay peninsula, both the rice mother and her child are represented by different sheaves of rice. On Bali and Lombock at harvest time, various bundles of rice sheaves are tied together, one representing a man and the other a woman, and are called 'husband and wife'. A similar notion is held in Upper Burma.

In North America, the Mandans and Mintarees thought that a certain 'old woman' who never dies made the crops grow and they gave this name both to the maize and those birds which they regarded as symbols of the fruit of the earth. To assure a good crop of rice, in some parts of India, the harvest goddess Gauri is represented at once by an unmarried girl as well as by a bundle of wild balsam.

Frazer concludes that there are grounds for regarding both Isis and her companion god Osiris as personifications of the corn. On the hypothesis suggested, Isis would be the old corn spirit, and Osiris would be the newer one, whose relationship to the old spirit was variously explained as that of brother, husband and son; for of course mythology would always be free to account for the co-existence.

According to M. A. Murray (1962):

The cults of Isis and Osiris are so inextricably mixed that it is impossible to disentangle them completely. The writing of her name shows that she was Queen, 'She of the Throne', and therefore naturally the partner of Osiris, the Occupier of the Throne. Her cult spread far and wide, so that there was a Thames-side temple of Isis in London and an altar to Isis at Chester. Her aspects are so many that she was known as the Myriad-named, but her chief epithets are 'Mother of God, Lady of Heaven'.

The cult of Osiris and Isis spread all over southern Europe, and into many parts of North Africa, and it continued to be a religious power in them until the close of the fourth century AD. At Philae, the worship of Osiris and Isis continued until the reign of Justinian and it only came to an end in Nubia because the emperor caused the temple to be closed by force and confiscated the revenues of the shrine. The ideas and beliefs which were the foundations of the cult were not even then destroyed, for they survived in Christianity. And the bulk of the masses in Egypt and Nubia who professed Christianity, transferred to Mary the Virgin the attributes of Isis the Everlasting Mother, and to the Babe Jesus, those of Horus. About the middle of the Ptolemaic Period, the attributes of Osiris were changed, and after his identification with Sarapis, i.e. Pluto, the god of death, his power and influence declined rapidly, for he

was no longer the god of life. In the final state of the cult of
Osiris and Isis, the former was the symbol of death and the
latter the symbol of life. (*Osiris,* vol. ii, pp. 305–06 by E. A. Wallis
Budge).

The Seclusion of Young Girls at Puberty

I cite at length and in detail (Frazer pp. 780–90) because this practice is so strange and bizarre that its widespread parallels can hardly
be attributed to independent origins (resulting, as it is alleged,
from inborn psychological tendencies). Up to now we have no
indication as to how and by what route this custom spread.

In New Ireland (in Pacific Melanesia), eye-witness accounts
have recorded that girls on their first menstruation are confined for four or five years in small cages, being kept in the dark
and not allowed to set foot on the ground. The atmosphere
inside is hot and stifling. When the doors are shut it is completely dark. There was only room for the girl to sit or lie down
in a crouched position. They are only allowed once a day to
bath in a dish or wooden bowl placed close to each cage. They
say that they perspire profusely. One of them was about four-
teen or fifteen years old and would soon come out; the other
two were about eight or ten and had to stay there for several
years longer.

Among the Ot-Damons of Borneo, on first menstruation,
girls of ages eight or ten are shut up in a little cell of the house
in almost total darkness and may not exit on any pretext
whatsoever. None of the family may see them and the confine-
ment often lasts seven years. Their body growth is stunted for
want of exercise. By the time they are finally brought out, their
complexion is pale and wax-like.

Among the Tlingit or Kolos Indians of Alaska, when a girl
showed signs of womanhood, she is confined in a little closed hut
or cage containing only a tiny air hole. In this dark and filthy
place she has to remain a year, without fire, exercise or
associates. Only her mother or a female slave might supply her
with food.

Among the Koniaks (Alaskan Esquimos), a girl at puberty was placed in a small hut in which she had to remain on her hands and feet for six months; then the hut was enlarged a little so as to allow her to straighten her back. In this posture she had to remain for six months more. All this time she was considered an unclean person with whom no one might associate.

Among the Guaranis of Southern Brazil, when symptoms of a girl's puberty first appeared, she was sewn up in her hammock, leaving only a small opening for ventilation. She was thus wrapped up and shrouded like a corpse for two to three days, or so long as the symptoms lasted, and had to observe a most rigorous fast. And a similar custom prevailed among the Chiriguenos of south-eastern Bolivia, except that the girl in the hammock was hoisted up to the roof for about three months. Some Indians of Guiana, after keeping the girl in her hammock at the top of the hut for a month, exposed her to certain large ants, which inflicted very painful bites. Sometimes, in addition, she had to fast, so that upon release she was near skeletal. Similar customs have been recorded from places as far removed from each other as New Guinea, Canada, Australia, North America, Cambodia, India and Bolivia.

Frazer's explanation for this practice is the alleged deeply ingrained dread which primitive people are supposed to have for menstrual blood. He maintains that they fear it at all times but especially on first appearance (p. 790). Such an explanation controverts certain facts. First although widespread, the custom is absent in numerous semi-civilised tribal societies (not to forget modern civilisation), and almost totally absent among culturally ultra-primitive food-gatherers, except those in close contact with food producers. This raises the question of why some tribal societies should ever have come to fear menstrual blood to begin with, since there can have been no real evidence to show that menstrual blood has any damaging or dangerous properties.

A striking example for this more natural viewpoint is found among the Mbuti pygmies of the Congo. Turnbull (1966:132) reports that the first appearance of menstrual blood in a young Mbuti girl is always a signal for rejoicing.

More on Frazer

The preceding examples from Frazer's *Golden Bough* were selected mainly for their strong historical associations and their wide geographical spread. This leaves hundreds of examples omitted here, of which we subsequently cite only a few more.

Judging Frazer's grand opus as a whole, we find him wholly wedded to a psychological explanation for identical worldwide cultural phenomena. Yet surprisingly, his main underlying motive, the priesthood of Nemi (a place near Rome devoted to the worship of the goddess Diana) seems rather extraneous to his thesis. Had he chosen Osiris as the main motivating force behind his examples, he would have stood on firmer ground.

To justify his choice of the 'King of the Wood' at Nemi as the guiding principle of his work, Frazer, in his Preface to *The Golden Bough* writes: 'The primary aim of this book is to explain the remarkable rule which regulated the succession of the priesthood of Diana at Aricia.' Frazer associated it with a widespread practice, summarised by him under the heading 'The Killing of the Divine King'. From the examples which will follow, it can be seen that the comparison is inadequate. As it happened, the priest or king of the wood at Nemi near Rome (Diana's sacred grove), was merely the guardian of a tree. Any stranger could challenge him and upon slaying him would replace him as 'king of the wood' (a king without a kingdom), only to experience a similar fate subsequently.

In contrast, the wider theme of 'killing the king' (as exemplified by Frazer himself), applied mainly to rulers or kings of whole countries. Only in this context are Frazer's examples applicable. His reasoning (which is only remotely related to the king or priest at Nemi) is as follows:

> We see a series of divine kings on whose life the fertility of man, of cattle and of vegetation is believed to depend, and who are put to death, whether in single combat or otherwise, in order that the divine spirit may be transmitted to

their successors in full vigour, uncontaminated by the weakness and decay of sickness or old age. Any such degeneration on the part of the king would, in the opinion of his worshippers, entail a corresponding degeneration of mankind, of cattle and the crops.

What we seem to have here then, unsaid by Frazer, is the Osiris motive. A revival of nature and with it the soil's fertility, which, according to myth, the separated parts of the murdered king Osiris bestowed on the land. Osiris's resurrection (whereby he became a god), can also be identified with the emergence of a rejuvenated king, who in his turn rejuvenated nature as the seasons demanded.

Some of Frazer's many examples of ageing kings being killed in this manner, and for the reasons given above, include the Kaffir Kingdom of Sofala; certain kings in parts of Cambodia; the Nubian god kings of Meroe, the Shilluk kings of the White Nile; the Central African kingdom of the Bunyora; the Jukun kings of West Africa; the kings of Angola, the Zulu king and many others. This shows that the practice extended widely over Asia and the rest of Africa.

The Sacrifice of Retainers at a King's Internment

This is another practice cited by Frazer which prevailed over long historical periods in many widespread areas. At a king's funeral his servants were buried alive along with him on the assumption that they would continue to serve him in his afterlife. This arose from early beliefs in a king's immortality. R. A. Jairazbhoy (1976:99/100) has compiled a short chronological list of examples. In the First Dynasty Egypt, at Abydos, funerary mounds of king and nobles were surrounded at burials by the immolation of artisans and servants who followed their master in death. At the Royal Graves at Ur (c. 2,650 BC) the number of sacrificed attendants ranged from 60 to 80. In China(by c. 677 BC) where the custom began in the Shang Dynasty, the victims numbered from e two to 30, but occasionally could reach 300.

In Japan (*c.* 2 BC), victims were buried alive up to their neck circling a ruler's grave and left to famish. After a lengthy abolishment, the practice was revived in the sixteenth century AD, with retainers committing harakiri. In India (*c.* 950 AD), 300 to 400 persons voluntarily sacrificed themselves. At the Maya Site at Palenque near the sarcophagus of the king, six skeletons were uncovered. In Mexico City at the death of the king Montezuma I (AD 1,469), many of his slaves joined him. Other authors, prominent among them Nigel Davies, have written about this practice at length.

As far as we know the practice originated in Egypt, together with a belief in the divinity and immortality of kings (divine kingship). From there it appears to have spread as a feature of early civilisation perhaps to Mesopotamia and other parts of Africa first, and later to India, China, Japan and the Americas.

Magic, Religion and Early Civilisation

Frazer impressively collates hundreds of examples of strange practices and superstitions and to bring order into the apparent chaos proposed an evolutionary line of origin. Magic was assigned a lower phase of human mental development, followed by the phases of religion and finally science. The underlying idea was based on the principle of 'unilineal cultural evolution' (comprehensively treated in Part Two). It is closely allied with Lamarckian heredity. Further, man's cruder superstitions were said to be due to the weaker intelligence of their originators, denizens of the lowest rung of the human evolutionary ladder. Such views about human development are now obsolete, a point which Frazer himself later conceded.

In his Preface to the 1922 edition of *The Golden Bough*, he expounds on his initial objective. Basing his work on circumstances surrounding the personality of the 'king of the wood' at Nemi, Frazer writes (1922:VII): 'Whether the explanation I have offered is correct or not must be left to the future to determine. I shall always be ready to abandon it if a better one can be suggested,' and 'I hope that after this explicit disclaimer

I shall no longer be taxed with embracing a system of mythology which I look upon not merely as false but as preposterous and absurd.'

It is possible that Frazer's conclusions are wanting on several counts. First with regard to so-called savages, and their supposed lower intelligence. All contemporary human groups, whatever their cultural status, are now recognised as biological equals, irrespective of their racial, ethnological or geographical origins. I further show in Part One that since his appearance on the evolutionary scene several hundreds of thousands of years ago, *Homo sapiens* has remained fairly static in his average brain size and most likely also in his intellectual powers. This view is supported even by Levi-Strauss. In a symposium 'Man the Hunter' (1960:351) he said: 'I see no reason why mankind should have waited until recent times to produce minds of the calibre of Plato and Einstein. Already two or three hundred thousand years ago there were probably men of similar capacity, who were of course not applying their intelligence to the solution of problems as do the more recent thinkers.'

Frazer's *Golden Bough* served also as a source book for Freud who used it to fortify his psychological ideas on human culture. Freud likewise maintains that there is progression from magic to religion and this in turn is succeeded by science. Jung went a stop further by maintaining that there is a deeper level of the personality, the collective unconscious, consisting of the collective myths and beliefs of the race to which the individual belongs, and which Jung termed 'universal archetypes' supposedly common to all humans (Macmillan, 1985:180). In this system of thought crude superstitions are described as the stirrings of an inherent primitive logic – sometimes denoted by psychologists of the Freudian school as expressions of 'primeval science'.

As contemporary studies of ultra-primitive food-gatherers and the reassessment of human mental evolution show, the process suggested by Freudian and Jungian psychology (and by Frazer as well, *see* Part Two), is in conflict with human experience. Firstly, the subject of science should have been entirely omitted from this sequence since it belongs to a type of phenomenon which allows no comparison with supernatural concepts. The internal combustion engine has no relation to idol worship. Yet a skilled auto-mechanic somewhere in Africa can at the same time, be the most ardent idol worshiper. Secondly, far from being the originators of crude superstitions, including magic, sorcery and other bizarre beliefs, the culturally most primitive people on this globe were, until recent times, the very ones bereft of this type of idolatry. Thus the twa-Mbuti of the Congo (Turnbull, 1966) and similar rare hunter-gatherer survivors show no traces of such beliefs.

Of the hundreds of examples of strange customs and superstitions catalogued by Frazer, none relates to ultra-primitive food-gatherers; the reason being that they did not possess such traits. The Frazerian examples are almost exclusively associated with food-producing populations as well as with those hunter-gatherer groups, who living in their proximity, became at least partly indoctrinated by their food-producing neighbours. A list of some of these latter groups, whom conventional ethnology still categorises as hunter-gatherers, have been recorded by anthropologists dating from the middle of the last century up to several decades ago. They include, and I repeat: the Veddahs of Ceylon; Andamanese aborigines, the Sakai; the Semang (a proto-Malay tribe); certain Negrito tribes of the Philippines; the Arunta and other Aborigine tribes of Australia. In the Americas, they include such tribes as the Dene, Salish, Paiute and the Indian tribes of California; further south there are numerous Brazilian tribes, while in the Chilean parts there still linger the Fuegians and their close neighbours, the Chonoans and the Onas; in the arctic regions we encounter various Eskimo tribes. Most of the above mentioned can only nominally be considered as hunter-gatherer people.

One may ask how did ultra-primitive food-gatherers acquire such belief systems from their food-producing neighbours in the first place? I have shown that such ideas cannot have resulted from any inborn evolution-related psychology. Consequently we are compelled to seek historical explanations. This is inadvertently implied by Frazer himself insofar as all the examples of beliefs and practices he lists in his work relate to food-producing communities with the chronologically earliest examples coming from Egypt and the Near East. As far as we know examples from food-producers of a lower cultural status (i.e. semi-civilised tribal societies) are of a later date.

24

Postscript to Frazer

I believe that I have submitted sufficient evidence to show that most of the worldwide cultural parallels cited by Frazer indicate common historical origins, despite our not being able to prove how the cultural transmissions were effected. Nobody can today dispute that tens, or even hundreds of thousands of years ago, groups of *Homo erectus*, and later *Homo sapiens*, found their way to some of the remotest corners of the then inhabitable globe.

Yet no one questions that both species, *erectus* and *sapiens*, originated in specific areas of the world, before spreading elsewhere. How those distant forefathers of ours managed to reach all these remote places, many of them seemingly inaccessible, has so far remained unexplained. Yet the fact of their spread is indisputable and must have included the crossing of large expanses of water.

The early dispersal of the human species (as the evidence of stone tools shows) must also have gone hand in hand with cultural spread, demonstrating that biological dispersal was linked with cultural spread. The subsequent diffusion of other, more complicated cultural traits can be deemed to have proceeded similarly. Thus we can assume that if we find cultural parallels, many of them very complex, in distant parts of the world, that they had common origins, irrespective of the method of their migration, whether known, merely implied, or unknown. I must, however, add that later-date cultural spread also involved the acculturation of already existing, less complex, cultures and populations by outsiders, involving colonisation, conquest and indoctrination.

The alternative view that identical cultural parallels in their worldwide context emerged independently without outside stimulus is hardly acceptable. And just as it is inadmissible to deny the common origin of the widely dispersed races of modern humans (an indisputable fact), so it is equally inadmissible to substitute a psychological evolutionary explanation for the independent origin of widely separated culturally identical parallels. The latter may have seemed plausible at a time when proof of the true antiquity and wide dispersal of hominids was still hidden in the ground, awaiting the spade of the archaeologist. Now that we are more knowledgeable in this respect, such a theory becomes mere conjecture. This is not denying, however, that rare exceptions do exist, although they cannot be attributed to a cultural evolutionary process.

Yet, independent duplicate development, inventions or discoveries, seen from a theoretical viewpoint, have always derived their main support from the postulate of a cultural evolutionary process, supposed to go parallel with biological evolution. Its successive stages, allegedly cause human culture to rise from the most primitive beginnings to high civilisation.

The evolutionary concept in culture (as repeatedly mentioned) is based on the biologically linked factor of the equal working of the human mind advanced by Freud. It implies that all human groups pass through the same cultural evolutionary stages everywhere, creating and duplicating quite independently the requisites of human culture.

If this principle were applicable, it would have meant that throughout the entire period of *Homo sapiens*'s presence, extending over about 300,000 years, cultural progress in all spheres of life would have become universally and uniformly distributed throughout the world. Instead we find that even in our modern world all stages of cultural development still co-exist side by side; hunter-gatherers at a near stone-age level; a potpourri of semi-civilised tribal societies retaining a multitude of bizarre customs; many under-developed civilised societies striving for higher Western standards; while our own materially advancing Western civilisation at the 'peak' of rationalism, science and technology, still carries the germs of past ethnic, racial, religious, nationalistic, economic and ideological conflicts.

What occurred in each case was that knowledge of the previous steps had been geographically disseminated, and it was this fact alone that made the next step possible, whether it was taken by one or several persons or groups. As pointed out before, Western civilisation is not a product of evolution or any other naturally promoted deterministic process, but has resulted from a series of historical coincidences, and in particular from the spread of new knowledge, assisted by Columbus's discovery of America, and the consequent opening up of the New World.

Turning to earlier history and prehistory we find the same uneven and varied cultural medley co-existing side by side. The numerous preceding examples of Frazer and other authors have shown that most cultural innovations can be traced back to distinct sources of origin from where they spread by direct or indirect contact. If such inter-relationships, historical or otherwise cannot be proved, we can, with few exceptions, confidently postulate them, whether they involve single cultural elements, whole cultures, or, in the last instance, entire civilisations.

25

Grafton Elliot Smith

A Tribute to a Presumed Ethnological Heretic

While Elliot Smith's preoccupation with Egyptian origins of the world's civilisations marked him, in the eyes of his adversaries, as an Egypto-Centrist and Hyper-Diffusionist, it can now be seen (in the light of more recent biological and archaeological discoveries) that the redeeming features of his ideas far outweigh his alleged shortcomings. Nevertheless, prejudices against him largely prevail. Thus Professor Sorenson advised me (personal communications): 'Why try so hard to resurrect Elliot Smith? Your thrusting him forward can only harm your possibility to persuade colleagues towards diffusion on a rational basis, for they have judged him already (ignorantly and with bias, it is true) and found him guilty.'

In spite of this, I have decided to take up Elliot Smith's case. It may come as a surprise to those who are not acquainted with his work that many of the basic facts relating to cultural dispersal elucidated in the present treatise had already been recognised and evaluated almost 80 years ago in his work, *The Migration of Peoples and the Spread of Certain Customs and Beliefs* (Manchester, 1915). George Carter, who has spent most of his academic career advancing similar ideas, found Elliot Smith a revelation when he came belatedly across his work. Carter discovered that:

> The man whom he and all anthropologists of his day had been taught to consider a mad man, was actually a witty, urbane, and above all, a well-informed scholar. His judgements of cultural events, were exciting and obviously the product of well organised research. His comments on Egyptian influences in America now begin to look prophetic. Those who have never read him had better, for he has been misrepresented badly.

Quoted below is a section of the report of the 'Symposium on Elliot Smith' held in London 1972. When Professor Lord Zuckermann introduced it, he called him a leading world authority who had rendered to anatomy what Lord Rutherford had given to physics. By a remarkable coincidence, both scientists were contemporaries and both had come from Australia to Cambridge. Elliot Smith arrived there from Sydney University in 1896 to do anatomical research at St John's College. His subsequent career began with an appointment in Cairo which put him, so to speak, into the lap of the world's most remarkable ancient civilisation. Anatomy led him to an examination of Egyptian skeletal remains, and mummies in particular. His studies, stretching over ten years, aroused his interest into all aspects of Pharaonic civilisation and in its continuation into that of the phenomenon of world civilisation as a whole. The catalyst to his ideas was mummification. The fact that mummified remains which incorporated the elaborate embalming technique, developed in Egypt over the centuries had been found in many parts of the world, suggested an Egyptian origin. In the process of further studies he became finally convinced that the world's ancient civilisations had been inter-related and that they had emerged from a historical process of continuity and diffusion. His views were further strengthened by the discovery of the survival of primitive food-gathering stone-age peoples, into the present age. For Elliot Smith, the whole ancient world, all human history, had now become one, and he became so firmly convinced of his views, and advanced them with such eloquent vigour, that his academic opponents, who could not discern the wood from the trees felt themselves mocked and offended.

Elliot Smith had suggested that it was only around 900 BC in Egypt that the complex structure, typified as 'heliolithic culture' had matured, while its roots reached back into the early pyramid age of 2,600 BC and before. It expanded by the addition of numerous accidental acquirements from neighbouring countries. The great migration of the 'heliolithic' (i.e. sun-worhip related) culture complex probably began shortly before, with its influence perceptible throughout the Mediterranean and Europe in the West, and parts of Western Asia in the East, eventually intruding into the Americas. (According to recent evidence advanced by R. A. Jairazbhoy, cultural spread from Egypt reached America as early as 1,200 BC.)

When Elliot Smith's original memoir on the subject was presented to the Philosophical Society of Manchester in 1915, his sole object had been to put together the scattered evidence supplied by the practice of mummification and other associated customs, in support of the fact that the influence of ancient Egyptian civilisation or a particular phase of it, had spread worldwide.

Since the theory of a 'heliolithic' culture and its spread was advanced by Elliot Smith more than 80 years ago, an enormous amount of evidence has been added, pre-dating his own findings by hundreds of years, indicating both trans-Pacific and trans-Atlantic intrusions into America, with many of the latter reminiscent of Egyptian elements. And although these new discoveries do not match Elliot Smith's earlier assumptions in detail (some of which had to be discarded), they more or less confirm his overall picture of a worldwide cultural process of diffusion and the theoretical assumptions cited in support.

As a matter of fact, now that the theory of cultural evolution, and the psychological argument of a universal symbolism attached to it, has lost its credibility, Elliot Smith's basic assumptions put forward as early as 1915, have gained a renewed and powerful prominence.

Elliot Smith had written then (1915!):

What arouses particular surprise is not so much the total lack of evidence to justify the amazing assumptions of the psychoanalytical school, as the fact that the conception of a universal symbolism is in accord with the essential principle of Freud's psychological method. The constant appeal to the meaningless phrase 'the similarity of the working of the human mind', for if any sense whatever is to be attached to this phrase it implies that man is endowed with instincts of much more complex and highly specialized kind than any insect or bird – instincts moreover which impel a group of men to perform at the same epoch a very large series of peculiarly complex, meaningless and fantastic acts that have no possible relationship to the 'struggle for existence' which is supposed to be responsible for the fashioning of instincts, as though it were a magical incantation against logical induction, and harping on the so-called 'psychological' argument, which is directly opposed to the teaching of psychology, are the only excuses one can obtain from the 'orthodox' ethnologist for his obstinate refusal to face the issue. The chief cause of confusion in recent times has been the repeated misuse of analogies and the failure to realize that the biological terms 'evolution' and 'convergence', cannot be applied to Human (i.e., cultural) phenomena in the way so many ethnologists are doing.

Now, 80 years later, one must regretfully admit that the principle of cultural evolution with all its attendant irrelevancies is still firmly entrenched in cultural theory. This despite the fact that since Elliot Smith's time an enormous body of new evidence has been added to indicate its obsolescence. Unfortunately, the theory of cultural evolution is not only still being upheld and taught by conventional contemporary anthropologists but is uncritically accepted by prominent scientists and thinkers. Among them are such luminaries as the late philosopher Karl Popper, the great physicist Stephen Hawking, and the late mathematician and philosopher Bertrand Russell, to name a few.

26

Postscript to Part Three

The task of unravelling the common origins of the world's cultural heritage and making a case for their interdependence, involves considerable difficulties. On the surface it appears simpler to attribute cultural origins to independent local developments, basing them (as has been done since Herbert Spencer's and E. B. Tylor's times) on tenets of cultural evolution. However, as preceding evidence reveals, cultural evolutionary theory (whether unilineal or multilineal) cannot be substantiated or supported either on historical, biological, or on any other purely theoretical ground.

On the other hand recent history documents the unique origin and spread of entire culture complexes, and of religious as well as that of ideological systems. The most conspicuous examples of the latter are Western civilisation (including Ancient Greece and Rome), the great religions of Buddhism, Christianity and Islam, and Marxist ideology. They all have spread worldwide with no indication of their having been independently duplicated anywhere. Neither is there any trace of an automatic cultural evolutionary process discoverable in their developments and spread.

What is more, we are virtually eyewitnesses of these fairly recent examples of cultural spread or diffusion. Yet the further we reach back into history, and more so into prehistory, the

scantier becomes the availability of evidence for cultural link-
ages and transmissions. In attempting to substantiate the claim,
for instance, that old-world influences were responsible for the
origins of the pre-Columbian American civilisations (and there
are many attested contacts), we unfortunately find that the
archaeological evidence for physical contact presently available
is still insufficient to prove this beyond reasonable doubt.
Establishing the derivation of Sumerian civilisation from Egypt,
or vice versa, faces the same degree of uncertainty.

Therefore, in order to incorporate earlier history and especially prehistory into our cultural model we have to rely on
circumstantial evidence. Here we can follow the precedent of
Charles Darwin, whose theory of the origin of species is likewise
largely based on circumstantial evidence.

Darwin replaced the thesis that each animal species had been
independently created by an act of god with his theory of the
common origin of all animal life, basing his argument mainly
on theoretical biological considerations. This has now found
general acceptance despite its failing to prove in many cases
a direct inter-relationship between widely separated animal
species.

Darwin wrote (1859):

> Turning to geographical distribution, the difficulties en-
> countered on the theory of descent with modification are
> serious enough. All the individuals of the same species,
> and all the species of the same genus, or even higher
> groups, are descended from common parents; and there-
> fore, in however distant and isolated parts of the world
> they may now be found, they must in the course of
> successive generations have travelled from some one point
> to all the others. We are often wholly unable to conjecture
> how this could have been effected.

It may be presumptuous to compare human cultural spread or
diffusion to Darwin's descent by modification in animals, for
one is a historical process and the other a biological one. But
the parallels are too striking to be ignored. Darwin had become

convinced that all living forms had a common ancestry; and also that there were many geographical and chronological gaps with no evidence to show how they could have been bridged. He solved the problem by postulating that all living forms must be inter-related, basing this thesis on well argued circumstantial evidence, which has now been substantiated by genetical science.

The diffusionist theory proposing worldwide cultural interrelationships faces similar problems. Thus, in cases of identical cultural expressions separated geographically or chronologically, with no material evidence to connect them, we also have to rely on the theoretical arguments submitted in this paper in order to postulate common links, because where evidence exists (and there are countless examples) they show that cultural elements or whole cultures and even entire civilisations, in different parts of the world, having similar features, are not the independent creations of their local inhabitants, but are derived from other similar cultures with whom they came in contact. The examples of Sir James Frazer (though not his theories) are perhaps the most eloquent witness to this process.

In summing up I can say this: while the diffusionist cultural theory proposed here, until perhaps substantiated by more extensive evidence of physical contact, remains vulnerable to criticism; its opposite, the theory, or myth, of cultural and social evolutionism (i.e. the independent invention theory) cannot be substantiated on any grounds.

Epilogue

Current history shows a tendency towards political world disorder. Presently no effective measures are in sight to heal the festering wounds caused by national, racial, religious, and above all of ethnically inspired dissension.

Note the examples of simmering discords in Rwanda/Burundi, Bosnia, Somalia, Afghanistan; the Catholic-Protestant tensions in Northern Ireland; the schism between Israel and the Arab World; the oppressive military dictatorships in many parts of the African continent, not to forget Burma; the rogue regimes of Iraq, Iran and Libya, prime promoters of world terrorism. Looming in East Asia is the monolithic communist regime of North Korea. Another similar enigma is presented by post-Maoist Communist China. Also, the world is still faced by political uncertainties, prevalent in post-nascent Russia.

And while the danger of a third devastating world war is presently held at bay, the future threat of a nuclear holocaust casts a permanent shadow on humanity's long-term survival.

The question is, how can the gulf between the world's many conflict groups be bridged?

A study of human nature based on biological principles (being part of the present book), forces us to recognise that the social animal *Homo sapiens*, presenting today's sole hominid survivor, is the product of an evolutionary process stretching over the best part of four million years.

In the process of this evolution, humans have become endowed with a genetic equipment which is universally applicable and which excludes any distinction based on racial or ethnic considerations. All present dissensions are the products of a historical process based on fallacious concepts of exclusivity. They are promoted by pseudo-science and educational misinformation.

What humans all over the world must be made aware of is that, whatever their racial and other diverse origins, they are basically the same. They inherit the same tendency for love and companionship as well as that of hatred and xenophobia. Thus, theoretically at least, a human universal brotherhood is possible. The alternative is war and destruction.

To solve the human dilemma and start the healing process, attempts must be made to resolve the conflict situations which plague humanity with the principle aim of extirpating the prejudices which divide human societies.

However, all this may become an idle dream as the Damoclean sword of nuclear annihilation hovers precariously over our heads. Thus, after an evolutionary journey stretching over four million years the very existence of *Homo sapiens* the sole hominid survivor, is now imperilled.

That the nuclear threat to humankind is now more potent than ever has been stressed by Nobel Peace Laureate Dr Joseph Rotblatt (a participant in USA's Manhattan Atomic Development Project). He said, 'there exist presently thousands of nuclear devices in the world ready to blast off at a moment's notice, each of them being of a capacity at least 20 to 100 times that of the atom bomb which destroyed Hiroshima.' In the face of these threats US atomic scientists have constructed a clock, which now clicks dangerously nearer towards 12 o'clock midnight doomsday.

In the present world political power game, the suggested limited NATO eastern expansion, which is clearly directed against Russia, is a further nail in the coffin of hope to solve the nuclear dilemma. Because, Russia with its military forces in disarray, can put her trust as a defence potential solely in the maintenance of still retaining its nuclear arsenal.

What then is the solution. Many prominent military observers in the USA believe that it is to invite Russia to become an integral part of the NATO alliance. This will bring the formidable western nuclear potential and the equally powerful Russian nuclear potential under a common military umbrella of control, as a first step towards the total elimination of nuclear armaments. Thereby, the former military NATO defence alliance, (having outlived its usefulness as a defence shield against the former Warsaw Pact), by including Russia and its former satellites, will now become a co-operative military and political alliance. Stretching from America's eastern Pacific shore via the American continent, via the Atlantic ocean, across Europe and Asia, to Vladivostok on the western shore of the Pacific, it almost encircles the northern part of the globe.

This act of political and military co-operation could become a unifying factor of global significance, affording an opportunity for solving all the other problems which plague humanity in future. Until this is achieved humankind will remain on the brink of nuclear annihilation.

Bibliography for Part Three

Beirne, D. R., 1971, 'Adze Hafting in the old and New Worlds', *Man Across the Sea*, Riley *et al.*, Texas University Press, USA

Carter, G. F., 1953, *Plants Across the Pacific*, W. M. Smith

— 1971, 'Pre-Columbian Chicken in America,' in *Man Across the Sea*, Riley *et al*, Texas University Press, USA

— 1973, A Hypothesis Suggesting a Single Origin of Agriculture,' Congress Report of Anthropological and Ethnological Sciences, Chicago, USA

— 1975, 'I am No Longer a "Standard Americanist"' *Anthropological Journal of Canada*, vol. 13, pp. 1–8

Coe, Michael, 1968, *America's First Civilizations*, New York, USA

Cowan, H. K. J., 1974, In *Current Anthropology*, XV/3

Darwin, Charles, 1859, *The Origin of Species*

Delson, E., 1984, *Ancestors, the Hard Evidence*, Alan R. Liss, New York

Daniel, Glyn, 1964, *The Idea of Prehistory*, Penguin Books

— 1971, *The First Civilizations*, Penguin Books

Davies, N., 1979, *Voyagers to the New World*, Macmillan

Dyer, T. A., 1890, In 'Nature'

Eckholm F. G., 1969, 'Transpacific Contacts,' in *Prehistoric Man in the New World*, Jennings and Norbeck, USA

Erman, 1927, *The Literature of the Ancient Egyptians*, Tubingen, Germany

Fell, Barry, 1986, in *Epigraphic Society Occasional Publications*, San Diego, California, USA

Frazer, Donald, 1965, *Theoretical Issues in the Transpacific Diffusion Controversy*, Sor. 32/452, USA

Frazer, James, Sir, 1960, *The Golden Bough*, Macmillan first published 1896)

Freud, Sigmund, *see* Jones

Griffith, J. G., 1966, *The Origin of Osiris*, Bruno Hessling, Berlin, Germany

Hatt, G., 1954, *Asiatic Motifs in American Folklore*, Underwood, Oxford

Heine, Geldern, R. von, 1974, 'American Metallurgy and the Old World,' N.

Barnard – article

Jairazbhoy, R. A., 1974, *Ancient Egyptians and Chinese in America*, London

— 1976, *Asians in Pre-Columbian America*

— 1979, in *Historical Diffusionism*, no. 27

— 1981, *Ancient Egyptians in Middle and South America*

— 1982, *The Spread of Ancient Civilizations*

Jones, E., 1961, The Life of Sigmund Freud, Pelican

J. Grun, Ruth, 1988, 'Linguistic Evidence,' in Support of the Coastal Route of Earliest Entry into the New World,' in *Man* vol. 23, no. 1

Kelley, D. H., 1971, 'Diffusion; Evidence and Process,' in *Man Across the Sea*, University of Texas Press, USA

Kraus, G., 1975, G. Elliot Smith (and W. J. Perry) *On Trial, New Diffusionist Offprint*, no. 2

Kroeber, A. L., 1948, *Anthropology*, 2nd edition, Harcourt Brace, New York

Lin Shun Sheng, 1956, 'Human Figures with Protruding Tongues', in *Bulletin of Institute of Ethnology*, no. 2, Taipei, Taiwan

Macmillan's Student Encyclopedia of Sociology, 1983

Meggers, B. J., Evans, C. and Estrada E., 1967, *The Early Formative Period of Coastal Ecuador*, USA

Magee, Bryan, 1986, *Popper; Fontana Modern Masters*

Mellaart, H. J., 1967, *Catal Huyuk*, Thames & Hudson

Murray, Margaret, 1962, *De I side de Osiris*

Needham, J. and Lu Gwei-Djen, 1985, *Trans-Pacific Echoes and Resonances; Listening Once Again*, World Scientific Publishers, Singapore

Ogburn, W. F., 1923, *Social Change*

Raglan, Lord, 1939, *How came Civilization*, Methuen

Riley, C. L. *et al*, 1971, *Man Across the Sea*, Texas University Press, USA

Sahagun, 1946, *Historia de los Cosas de Nueva Espana*, Mexico

Sertima, Ivan Van, 1987, 'African Presence In Early America', in *The Journal of African Civilizations*, New Jersey, USA

Slobodin, R., 1977 *W. H. R. Rivers*

Smith, G. Elliot, 1915, *Memoirs and Proceedings of the Manchester Literary and Philosophical Society*, vol. 59, part 2

— 1916, *The Influence of Ancient Egyptian Civilization in the East and America*, John Ryland's Library

— 1924, *Elephants and Ethnologists*, Manchester University Press

— 1929 (1915 reprint), *The Migration of Early Culture*, Manchester University Press

Sorenson, J. L., 1971, *Annotated Bibliography*, F. A. R. M. S., Utah, USA

Soustelle, J., 1985, *The Olmecs*, Doubleday, New York

Strauss, C. Levi- 1968, *Structural Anthropology*, Allen Lane

Tylor, E. B., 1881, *Anthropology*

Thorne, M. H. *et al*, 1984, *Modern Homo Sapiens Origins*, (*see* Delson)

Tolstoy, P., 1963, 'The Manufacture of Bark-cloth', in *Tynas* 25/646 USA

Turnbull, D. C., 1966, *Wayward Servants*, Eyre & Spottiswood

Washburn, 1968, in Symposium, 'Man the Hunter', USA

Glossary

Acheulian – The tradition of tool-making distinguished by hand-axes, with a duration of from *c.* $1^{1}/_{2}$ million years ago to *c.* 30,000 BP.

Alleles – Genes, which can mutate on to another, and which occupy the same relative position (locus) on homologous chromosomes, and which undergo paring during meiosis.

Artefact – Any object fashioned for use by man.

Asymmetrical – Having the two sides unlike.

Aurignacian – Major Upper Palaeolithic tradition of Europe from *c.* 34,000 years BP to *c.* 27,000 BP, characterised by long retouched blades, scrapers and split-base bone points (see also Chapter 5, Part 1).

Australopithecus (A.) – The earliest defined genus of hominid. (Almost all specimens occur in the time-range between 4–1.3 million years ago.) The following species have been named:

  A. afarensis – early form in Ethiopis.

  A. africanus – early form in Southern Africa.

  A. robustus – later robustus form from S. Africa.

  A. boisei – formerly Zinanthropus, also S. Africa.

Bantu – East Africans who speak languages of a single stock, so named.

Bipedalism – The habit of walking on two feet.

Blade – A parallel-sided blade of flint of nearly uniform thickness.

Brain – Wernike area: controls speech.

Broca area – Controls the muscular movements for sound.

Burin – A flint chisel.

Bushmen – Generally referring to the Khoi-San, aboriginal hunter-gatherers of South Africa considered by some a derogatory term.

Cenozoic – Same as Tertiary.

Cerbellum – Part of the brain projecting backwards.

Cerebellum – A special part of the brain projecting at back – 'little brain'.

Cerebral hemispheres – The different parts of the brain.

Cerebrum – Front part of the brain.

Chopper – A crude core of flint or other stone, sharpened by retouching along a single edge.

Chopping tool – The same as a chopper, but retouched by striking alternative blows from either side.

Chromosomes – Thread-shaped bodies, numbers of which occur in every nucleus of animal or plant cells.

Civilisation – Sometimes defined as a Post-Neolithic, complex society, mostly connected with a first appearance of kings and king-related gods, organised religion, monumental architecture, writing and a calendar.

Clactonian – An early flake industry from south-east England.

Cladogenesis – The branching off of evolutionary variations from the main-stem.

Cleaver – A form of biface tool found alongside the hand-axe on Acheulian sites especially in Africa.

Core (flint) – The central portion or nucleus of a flint nodule, formed by the removal of flakes or flake-blades.

Cortex – Outer part or rind; the grey matter of the brain.

Cultural evolution – The belief that human culture progresses automatically from a low cultural level to one of greater complexity. Originally considered to be a continuation of biological evolution.

Darwinism – The transmutation of species by natural selection (better called natural elimination). Darwin accepted part of Lamarckian heredity.

Dimorphic – Having two distinct forms.

Encephalisation – Enlargement of the brain substance.

Encephalon – The brain.

Endocast – Inner shape of the brain case, shown on a cast.

Evolution – The biological change in species. In Darwin's terms: 'descent by modification'.

Flake – A piece of stone, or flake, struck from a core.

Fossil – A relic or trace of former living organisms left in rock formations.

Gene – A protein particle, forming part of a chromosome, by means of which hereditary character are transmitted.

Glacial period – A time span encompassed by any one of the four major glacial advances of the Pleistocene.

Gravetian – Upper Palaeolithic tradition of Europe, extending from Russia to Spain and lasting from *c.* 27,000 BC to *c.* 19,000 BC.

Hand-axe – An almond-shaped stone implement mostly of flint or quartzite, symmetrical both bilaterally and bifacially and retouched on both borders.

Hominids – The family of man, including all species of *Australopithecus* and *Homo*.

***Homo* species** – In their succession: *Homo habilis, Homo erectus, Homo sapiens*. (*Homo sapiens neanderthalensis* is now considered a sub-species of *H. sapiens*.)

Hybrid – The offspring of the union of two distinct species.

Intelligence – Mental brightness, a person's degrees of knowledge.

Kenyapithecus – Considered an early ancestor of the hominid line. Same as *Ramapithecus*.

Knapper – a stone tool-maker.

Lamarckism – The theory of heredity which proclaims that characteristics acquired during a lifetime, or by the greater use or disuse of organs, are inheritable.

Levalloisian-tools – A primitive hand-axe; forerunner of the Acheulian type.

Lingula – A small, bottom dwelling shellfish – an early living form surviving into the present age (*Phylum Brachiopoda*).

Littoral – Inhabiting the seashore.

Lobe – A division of the brain frontal lobe: controls movement back or occipital lobe: controls vision and emotion parietal lobe: controls and integrates sensory input.

Locus – A particular position in a particular chromosome.

Magdalenian – Upper Palaeolithic tradition, mainly in Europe from *c.* 16,000 BC to *c.* 10,000 BC.

Mammalia – The highest class of animals producing living young and feeding them milk by means of their teats.

Melannin – A genetically linked granular pigment in human and animal tissue, causing a dark skin complexion.

Mesolithic – The pre-agricultural cultural period from *c.* 10,000 to 8,000 BC in the old world.

Microlith – A small stone artefact, normally under 3 cm, used for arrow tips and sickle blades.

Mousterian – A middle Palaeolithic cultural assemblage mainly attributable to Neanderthal man.

Mutation – Sudden change or variation in the genetic equipment of a germ cell.

Neanderthalers (Neanders) – *Homo sapiens neanderthalensis* (see under *Homo* species).

Neolithic (New Stone Age) – First appearance of polished stone tools. Mainly associated with the advent of agriculture and stock-rearing.

Neo-Darwinism – Darwinism minus Lamarckian heredity. Lamarckism became obsolete with the advent of genetics.

Neo-Lamarckism – Attempts to revive a Lamarckian-type inheritance.

Neurology – The study of nerves.

Obstetric limitation – The anatomical limits in the size of the woman's birth channel.

Oldowan – Adjective form of 'Olduvai', applied to the early pebble tools of south and east Africa, originally found at Olduvai Gorge in N. Tanzania.

Olduvai Gorge – In N. Tanzania, famous for its extensive deposits of early human fossils and stone-tools.

Osteodentokeratic – A term coined by Raymond Dart of an early hominid culture at Mapakansgate, South Africa, using bones for tools.

Palaeolithic – Old Stone Age, variously estimated: began *c.*

3 million years ago and finished *c.* 8,000 BC (the beginning of the Neolithic).

Palaeo-Neurology – A study of the 'early structure of the nervous system.

Pebble tool – A crude tool made by breaking or splitting a water-rounded pebble.

Phylogeny – History or development of a species, or race.

Pithecanthropus (Java man) – A *Homo erectus* fossil found in Java without stone-tool connections.

Pleistocene – Sixth subdivision of the Cenozoic era. Estimated to reach back over a million years, up to 8,000 years ago.

Polished stone axe – A form of stone axe, mostly made for hafting, produced by pecking, grinding and polishing. Estimated to have only appeared since the beginning of the Neolithic period.

Post-natal – Referring to the time just after a baby's birth.

Pressure flaking – Retouching flint implements by pressing the edge with a piece of wood or bone, instead of striking blows.

Primates – Include man, the apes, Old and New World monkeys, lemurs and tarsiers.

Punctational evolution – In contrast to a slow and gradual process, a more explosive form of evolutionary change.

Quern – A hollowed stone for grinding cereals.

Ramapithecus – Same as *Kenyapithecus*.

Reticulate evolution – Results from a network-like mating of different races within the same species.

Retouching – The process of sharpening flint implements by the removal of small chips.

Scraper – Class of stone tools with retouch along one or more edges.

Sinanthropus (Peking man) – A race of *H. erectus* whose remains were found in Choukoutien, near Peking.

Speciation – A division into different species.

Steinheim (female) – A *Homo sapiens* skull, *c.* 150,000 years old found near Steinheim, Germany.

Swanscombe Fossil – A fragmentary female *H. sapiens* skull, of *c.* 200,000 BP found in England.

Taxa – A specific class of living organism.

Taxonomy – Science of the classification of living organisms and their sub-divisions.

Tertiary – Same as Cenozonic. The geological period from *c.* 70 million years ago to the present.

Twa-Mbuti (Pygmies) – The type of aborigine in the Ituri Forest of the Congo.

Zinanthropus – same as *Australopithecus boisei*.